KU-104-918

FRONTISPIECE

*Color photograph of the Great Spiral in Andromeda;
48-inch Schmidt telescope, Palomar.*

(*California Institute of Technology, Pasadena, Cal.*)

THE STORY OF ASTRONOMY

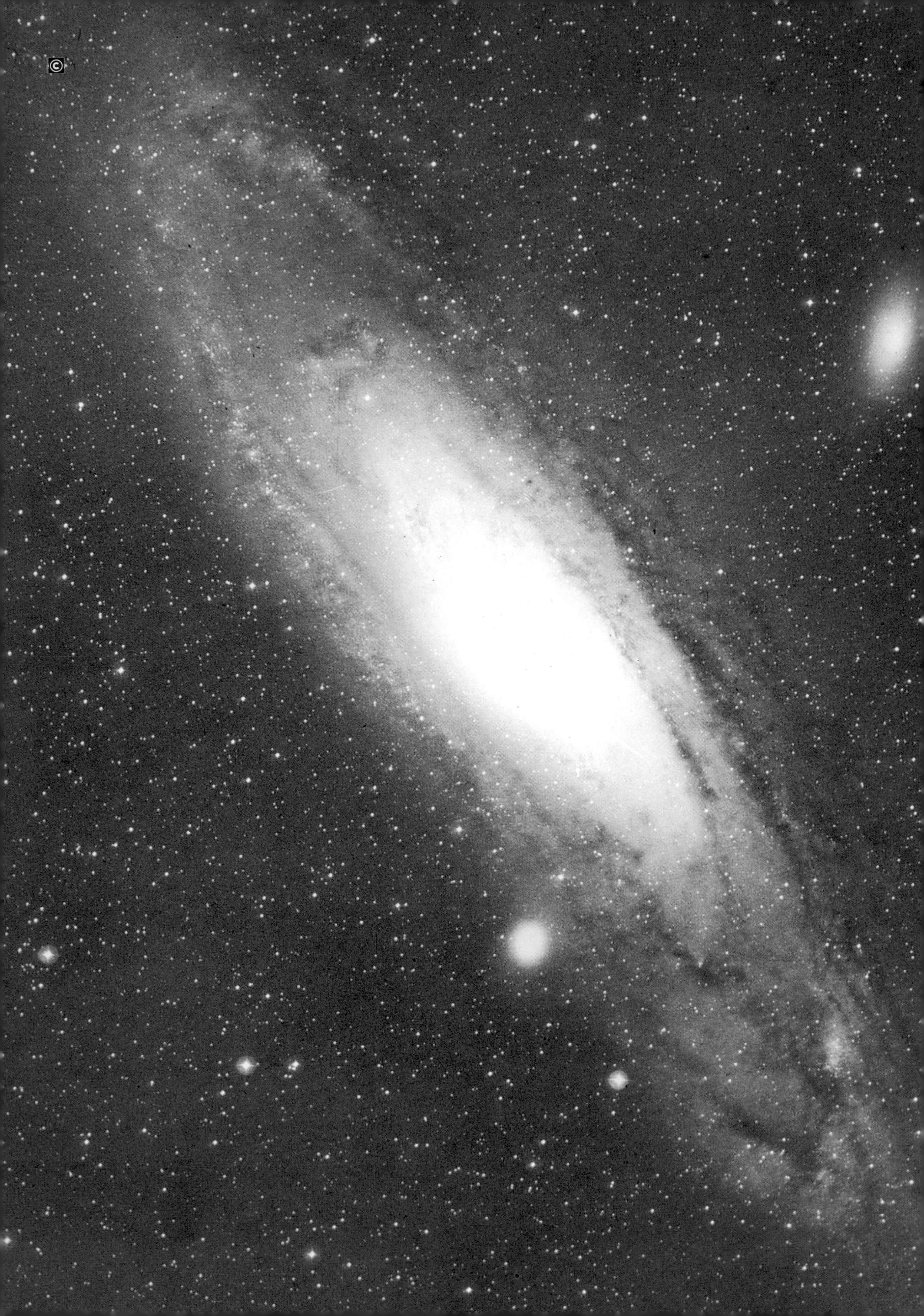

THE STORY OF

ASTRONOMY

PATRICK MOORE

BOOK CLUB ASSOCIATES
LONDON

This edition published 1973 by
Book Club Associates
By arrangement with Macdonald & Company (Publishers) Limited

Copyright © Macdonald & Company (Publishers) Limited, 1972

Reprinted 1974

Printed in Great Britain by
Morrison & Gibb Ltd, London and Edinburgh

contents

foreword

ASTRONOMY IS THE OLDEST science in the world. Today, it is also one of the sciences which is undergoing the most rapid development. Our ideas today are very different from those of 1900, 1940 or even 1960. Each year brings its quota of new discoveries and new surprises.

What I have tried to do, in this book, is to tell the story of astronomy, beginning in the remote past and taking it through to April 1972. Inevitably, much has been left out and much more has been glossed over; but I hope that what I have written may be of interest to those who read it.

My sincere thanks are due to the publishers for their help—in particular to Miss Susan Hodgart.

PATRICK MOORE

Selsey, April, 1972

1 the sky above us

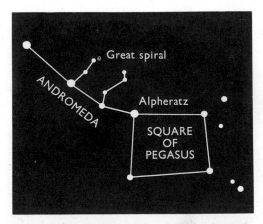

THE ANDROMEDA SPIRAL. *One of the nearest of the external galaxies, photographed with the 200-inch Hale reflector at Palomar Observatory. The two satellite galaxies are clearly shown*

POSITION OF THE ANDROMEDA SPIRAL. *The Spiral is just visible to the unaided eye on a clear night*

FAR AWAY IN SPACE, so remote that it looks like nothing more than a misty patch of light, lies an object which astronomers know as the Great Spiral. It is just visible without optical aid when the sky is really dark, and a pair of field-glasses will show it clearly.

The neighboring stars belong to the group or constellation known as Andromeda, and are members of our own stellar system. The Great Spiral is more distant than Andromeda; it is an independent star-system, and is so far away that its light, moving at 186,000 miles per second, takes over two million years to reach the Earth. When we look out into space, we also look backward in time; we are seeing the Spiral not as it is now, but as it used to be over two million years ago. It is a vast system made up of more than 100,000 million suns, many of which are a great deal larger and hotter than our own.

Ideas of this kind are difficult to grasp, and it is tempting to regard the Earth as the most important body in the universe. Nothing could be further from the truth. Every star is a sun; astronomical distances have to be reckoned in millions of millions of miles, and the Earth proves to be utterly insignificant.

The Earth is a typical *planet*, travelling round the Sun and completing one journey in 365¼ days. Of the eight other planets which move round the Sun at various distances, four are smaller than our world, and the remaining four much larger. Jupiter, the largest planet, is big enough to contain over 1,300 globes the size of the Earth. It is not solid and rocky; it is made up of dense gas, so that life there appears to be out of the question. Whereas the Earth has a single *satellite*, our familiar Moon, Jupiter has twelve.

The Sun, like all the other stars, sends out a tremendous quantity of light and heat. The Moon and the planets, however, are not self-luminous, and shine only because they reflect the Sun's rays. Seen from a distance of thousands of miles, the Earth would appear to shine in the same way.

Together with other bodies of lesser importance, the Sun, planets and satellites make up the *Solar System*. This system represents only a very small part of the universe, but is extremely large judged by everyday standards. The Moon, the Earth's nearest neighbor, is almost a quarter of a million miles away, while the distance of the Sun is 93 million miles. Pluto, the most remote of the nine planets in the Solar System, moves round the Sun at an average distance of well over 3,000 million miles.

This may be shown by means of a scale. Let us represent the Sun by a globe 2 feet across. The Earth will then be the size of a pea, placed at a distance of 215 feet; Jupiter will become a large orange at a distance of one-fifth of a mile, while Pluto will become a pin's head more than 2 miles away from the model Sun. On this scale the nearest star, known to astronomers as Proxima Centauri, will be several thousands of miles off. If the 2-foot Sun is set down in the middle of England, Proxima will have to be taken to Siberia or the United States in order to represent its distance correctly.

PLANETS	DIAMETER IN MILES	MEAN DISTANCE FROM THE SUN IN MILLIONS OF MILES
MERCURY	3,000	36
VENUS	7,700	67
EARTH	7,926	93
MARS	4,200	141·5
JUPITER	89,000 (Equatorial) 83,000 (Polar)	483
SATURN	75,000 (Equatorial) 69,000 (Polar) Rings 175,000	886
URANUS	29,300	1,783
NEPTUNE	31,500	2,793
PLUTO	3,700?	3,666

SEGMENT OF THE SUN 864,000 MILES IN DIAMETER

COMPARATIVE SIZES OF THE SUN AND PLANETS. *The sizes of the Sun and planets, drawn to the same scale.*
Drawing by D. A. Hardy

All the bodies in the sky seem to move in an east-to-west direction, giving the false impression that the Earth lies in the centre of the universe. This daily impression of movement has nothing to do with the stars or planets, and is due entirely to the fact that the Earth is rotating on its axis from west to east. The stars are not fixed in space, as ancient peoples believed; they move in various directions at great speeds, but are so far away that their individual motions cannot be detected except over long periods, and the patterns or *constellations* of stars seem to remain more or less unaltered. There is a good everyday comparison. A high-flying jet aircraft will seem to crawl across the sky, while a near-by bird will flash past quickly—yet in reality the jet is the faster of the two.

Though the stars are suns, they are not all alike. Some are hundreds of millions of miles in diameter, while others are no larger than the Earth; some are cool and red, others white and extremely hot. Some of the stars brighten and fade over short periods; some prove to be made up of 'twins' so close together that to the unaided eye they appear as one; some are surrounded by clouds of gas, and some are spinning round so quickly that they are shaped more like eggs than billiard-balls. It is possible, too, that many of them may have planet-families of their own, and that some of these planets are inhabited.

On a clear night it is possible to see about two thousand stars without the aid of a telescope, and the leading constellations are easy to find. Most people can identify the Great Bear, Orion and the Little Bear, while Australians and New Zealanders are just as familiar with the Ship and the Southern Cross. The powerful instruments used by modern astronomers show many additional stars, and it is now known that our stellar system or *Galaxy* contains about 100,000 million separate suns. Each star is using up energy as it shines, and there must come a time when the star will have used up all its supply of 'fuel', so that it will die. The Sun is no exception. It will last for thousands of millions of years yet, but it will not last for ever.

The bright stars visible at night are members of the Galaxy in which the Sun and the Earth lie. The Great Spiral is much more remote; it too contains suns of all types, and is a system larger than ours. The world's most powerful telescope is capable of showing 1,000 million galaxies.

Ancient peoples had little idea of the true nature of the universe. They had no telescopes, and had to rely solely upon their eyes. Today, large instruments are used for astronomical research; the biggest telescope so far in use, the reflector at Palomar in California, collects its light by means of a mirror 200 inches in diameter, which acts as a giant 'eye'. The Russians' main effort in the construction of large telescopes has been their 236-inch reflector, which was almost ready for testing by the early part of 1972.

If a stone is dropped into a calm pond, ripples will be formed. The distance from one crest to another is known as the wavelength, and there is a comparison here with light, which may be regarded as a wave-motion. The color of light depends upon its wavelength; red light has a wavelength longer than that of blue, and so on. Radiation of still longer wavelength cannot be seen at all, but it can be studied by means of special instruments known as

STAR-TRAILS. *A time-exposure was made with the camera pointing at the North Celestial Pole; the stars seemed to move slowly across the sky, so producing trails. This apparent movement of the sky is due to the real rotation of the Earth on its axis. Photograph by Allan Lanham*

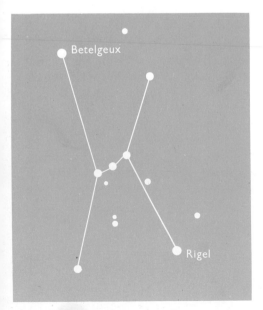

ORION. *One of the most brilliant constellations in the sky, with two first-magnitude stars, Betelgeux and Rigel*

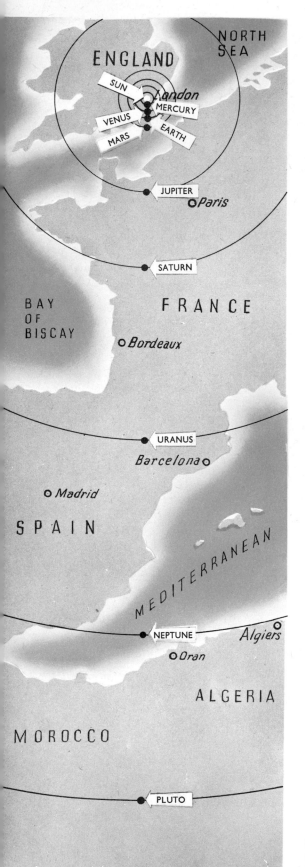

radio telescopes. The best-known of these radio telescopes is the British one at Jodrell Bank, Manchester, and takes the form of a wire dish 250 feet in diameter. It is used to collect radio waves coming from the depths of space, and is capable of detecting waves from distant galaxies which are particularly powerful in the radio range.

Since 1957, when the Russians launched the first Earth satellite, rocket astronomy has come very much to the fore. Its first achievement came with the pioneer American satellite, Explorer I, which detected the zone of intense radiation around the Earth which we now know as the Van Allen zone. In 1961 Yuri Gagarin made the first manned foray beyond the atmosphere; less than nine years later Neil Armstrong and Edwin Aldrin stepped out on to the surface of the Moon. Meantime, automatic probes had been dispatched to Venus and Mars, and plans were being made to send vehicles out to the frontiers of the Solar System. Progress during those few years was remarkable indeed. Astronomy had stepped out into space.

Yet astronomy itself is the oldest science in the world. Even the earliest men—the cave-dwellers of long ago—must have looked up at the skies and wondered just what the stars were; it was natural to worship the Sun and Moon as gods, and equally natural to believe the Earth to be a flat plain in the exact centre of the universe. Less than 600 years ago it was still believed that the Earth must be the most important body of all, and men who lived at the time of the Spanish Armada refused to believe that our world goes round the Sun. The apparent positions of the stars, and the movements of the planets in the sky, had been measured with great care, but their significance remained unknown.

In 1609 Galileo Galilei, an Italian Professor of Mathematics, first looked at the heavens through a telescope. It was he who studied the moons of Jupiter, the craters of the Moon, and the thousands of faint stars which make up the Milky Way. His telescope was low-powered judged by modern standards, but during succeeding years better instruments were built, and the first true observatories came into being. Since the time of Galileo and his primitive telescope, astronomical knowledge has increased steadily. Quite apart from our practical exploration of the Moon and planets by means of the new rockets, we have found out a great deal about the remote stars and star-systems. Yet the more we learn, the more we realize how much we still do not know.

It is impossible to understand and appreciate all this without some idea of what happened in past ages. The story of astronomy is more fascinating than any other branch of history, and by tracing its progress it is possible to gain a real understanding of the Earth, the stars and the universe itself.

SCALE OF THE SOLAR SYSTEM. *The distances of the various planets from the Sun are shown on this scale drawing*

THE HALE REFLECTOR. *The 200-inch reflector at Palomar Observatory, so far the largest telescope in the world*

2 watchers of the stars

TOTAL ECLIPSE OF THE SUN, 1919. *The drawing shows an extensive prominence which was nicknamed 'the Anteater Prominence'. It is very seldom that a prominence of this size is seen during a total eclipse. Drawing by D. A. Hardy*

IT IS IMPOSSIBLE to tell just when astronomy began. The cavemen of thousands of years ago must have looked up into the sky and marvelled at what they saw there, so that in one sense they were 'observers'; they noticed unusual happenings, such as eclipses, and written records of man's findings and theories go back to the dawn of history.

Early races believed the Earth to be flat and stationary, with the entire sky revolving round it once a day. The idea that the world is a globe nearly 8,000 miles in diameter, whirling round the Sun at a speed of some 66,000 m.p.h., did not occur to them. Some of their old ideas sound strange to our ears. The Vedic priests of India believed the Earth to be supported upon twelve massive pillars; during the hours of darkness the Sun passed underneath, somehow managing to pass between the pillars without hitting them. Even more peculiar was the Hindu theory, according to which the Earth stood on the back of four elephants; the elephants in turn rested upon the shell of a huge tortoise, while the tortoise itself was supported by a serpent floating in a limitless ocean.

Ancient man had to begin at the very beginning, and mistakes were inevitable, but at least useful observations could be made, and many of the early records have proved to be of tremendous value. Probably the first real students of the sky were the Chinese.

PARTIAL ECLIPSE OF THE SUN, *June 19, 1936, 16h 20m. From an observation made by Patrick Moore, using a 3-inch refractor, × 100. Three sunspots are shown. Drawing by D. A. Hardy*

TOTAL ECLIPSE OF THE SUN, *February 15, 1961. The line of totality extended across Southern Europe. The eclipse was shown on B.B.C. television from three stations in Europe in succession; St. Michel (France), Florence (Italy) and Mount Jastrebac (Jugoslavia). Pictures obtained from France and Italy were good; conditions in Jugoslavia, where the author was commenting, were affected by cloud. This photograph was taken off the television screen during transmission from France*

THEORY OF AN ECLIPSE OF THE SUN. *The Moon's shadow just reaches as far as the Earth; a partial eclipse is seen to either side of the belt of totality*

[*The drawing is not to scale*]

About 3000 B.C. they adopted a 'year' of 365 days, which enabled them to work out a calendar. It mattered little to them whether the Sun went round the Earth, or the Earth went round the Sun; the 365-day year was correct in either case. The Chinese Emperor's astronomers produced a reliable calendar, and were also able to tell when eclipses were due.

The Moon has no light of its own, and is a relatively small body only 2,160 miles in diameter. The Sun, with a diameter of 865,000 miles, is so much farther away than the Moon that it appears almost exactly the same size in the sky, and this is why 'solar eclipses' can take place.

During its monthly journey round the Earth, the Moon must sometimes pass in front of the Sun. At such times the dark side of the Moon is turned towards us, as shown in the diagram; and since this side does not shine, the Moon cannot be seen. As it moves between the Earth and the Sun, blocking out a part of the Sun's surface, a 'bite' appears in the edge of the solar disk, and this bite becomes larger as the eclipse progresses. If the Moon completely covers the Sun, the eclipse is total, and the solar atmosphere flashes into view, with startling effect. Normally it is impossible to see this solar atmosphere with the unaided eye or with ordinary telescopes, since it is overpowered by the brilliance of the Sun itself. When the Moon acts as a screen, however, the full spectacle becomes visible—the glorious pearly gas known as the *corona*, as well as the red *prominences* which rise from the solar surface. No total eclipse lasts for more than about eight minutes. As soon as part of the Sun's disk reappears, the prominences and the corona are blotted out; the Moon moves steadily in its path, and the eclipse comes to an end.

The Chinese had different ideas, and had no thought that the Moon could be concerned in any way. They believed that a dragon was trying to eat the Sun, and their remedy was to scare the beast away by making as much noise as possible. The whole population took part, shouting and wailing, and beating gongs and pans to add to the uproar.

However, they did know that any eclipse is likely to be followed by another eighteen years eleven days later, and by reckoning according to this period—the so-called *Saros*—they could avoid being taken unprepared. Now and again mistakes were made, and there is one famous story which relates how two luckless Court astronomers, Hsi and Ho, were executed because they had failed

Sun

Moon

Partial Eclipse

Total Eclipse

Eart

to predict an eclipse. This may be nothing more than a legend, but in any case there can be no doubt that eclipse forecasts were being made by men who lived 4,000 years ago.

A solar eclipse does not happen every month, because the Moon's path or *orbit* is appreciably tilted, and in most cases the dark and therefore invisible Moon passes either above or below the Sun in the sky, so that no eclipse occurs.

The Chinese did not confine their records to eclipses. They also recorded *comets*, which they regarded as unlucky. It is now known that a comet is made up of comparatively small particles surrounded by an envelope of gas, and is completely harmless, but the spectacle of a brilliant comet with shining head and long tail was enough to strike terror into the hearts of primitive peoples.

The Chinese and other early observers, notably the Egyptians, were content to compile their records without troubling greatly about what the various phenomena meant. This was probably because the heavenly bodies were regarded as divine, and for many centuries it was impossible to separate astronomy from

COMET AREND-ROLAND, 1957. *This was one of the two naked-eye comets seen in 1957; it was a fairly conspicuous object for some weeks during May, when it was in the northern part of the sky. Photograph by F. J. Acfield, Forest Hall Observatory, Northumberland*

VENUS AND HALLEY'S COMET, 1910. *When this photograph was taken, the comet was almost at its brightest, but Venus is necessarily over-exposed. Halley's Comet will next be brilliant in 1986. Photograph by H. E. Wood, Union Observatory, Johannesburg*

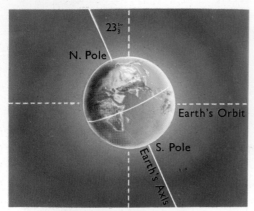

INCLINATION OF THE EARTH'S AXIS. *The equator is inclined by 23⅓ degrees to the plane of the Earth's orbit*

astrology—the superstition of the stars. Even today such confusion is not uncommon, but there is no excuse for it. Astronomy is an exact science, whereas astrology is of no value to anybody.

The stars appear to keep to the same patterns in the sky, while the much nearer Sun, Moon and the planets appear to wander slowly about. Yet the 'wanderers' do not move irregularly; they keep to a definite belt in the sky, known as the Zodiac. According to astrologers, a person's character and destiny are affected by the positions of the Sun and other bodies of the Solar System at the time of birth. The whole idea is completely baseless, but only during the past two or three centuries has astrology been finally discredited. Until then it was regarded as more important than true astronomy.

Apart from recording startling events such as eclipses and comets, as well as drawing up a workable calendar, the Chinese made little progress astronomically. The Egyptians, however, proved to be extremely skilful at measuring the apparent positions of the stars, and they 'lined up' their famous Great Pyramid in accordance with what was then the North Pole of the sky. This again is important, because it has given an excellent clue to the age of the Pyramid itself.

The Earth's axis of rotation is tilted at an angle of 23⅓ degrees to the perpendicular. At the present moment the axis points northward to a position close to the bright star Polaris, which is therefore our Pole Star. When the Pyramid was built, the pole of the sky was in a different position, and the Pole Star of those days was a much fainter object—Thuban in the constellation of the Dragon. The Earth is not a perfect sphere; it is slightly flattened, so that the equatorial zone bulges out, and the diameter measured through the poles is 26 miles shorter than the diameter measured through the equator. The Sun, Moon and other bodies exert a pull on this bulge, and the result is that the Earth's axis seems to wobble very slowly, in the manner of a top which is about to fall. This causes an effect known as *precession*. The polar point describes a circle in the sky, and has shifted considerably since the time when the Great Pyramid was set up.

Astronomy was developing. The stars were divided into definite constellations; observations began in other countries, and calendars were improved. Then, about 600 years before Christ, came the Greeks—and with them a revolution in scientific knowledge.

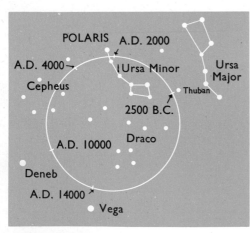

APPARENT MOVEMENT OF THE NORTH CELESTIAL POLE. *The shifting is due to precession effects, or the slow movement in the direction of the Earth's axis. In ancient times the Pole Star was Thuban in Draco; at present it is Polaris in Ursa Minor; by the year A.D. 14000 the northern polar star will be the brilliant Vega, in Lyra*

AN ECLIPSE OF THE MOON, *in four stages. Drawings by D. A. Hardy* ▶

3 the Greek astronomers

THREE PHOTOGRAPHS OF AN ECLIPSE OF THE MOON. *It is easy to see that the Earth's shadow on the lunar surface is curved*

THALES, FIRST OF THE great Greek astronomers, was born about the year 624 B.C. Like the Chinese and the Egyptians, he studied the stars, but he went further, and tried to explain what he saw. He may have realized that the Earth is a globe, though since all his original writings have been lost it is impossible to be certain.

The first definite arguments against the traditional flat-earth theory were advanced by Aristotle, who lived from about 384 to 325 B.C.

As Aristotle pointed out, the stars appear to alter in height above the horizon according to the observer's position on the Earth. From Greece, the Pole Star appears high above the horizon, because Greece is well to the north of the Earth's equator; from Egypt, the Pole Star is lower down, and from southern latitudes it can never be seen at all, since it never rises. On the other hand Canopus, a brilliant yellowish star in the southern part of the sky, can be seen from Egypt but not from Greece. This is easy to explain on the assumption that the Earth is a globe, but cannot be accounted for by supposing the Earth to be flat. Aristotle also noticed that during a lunar eclipse, when the Earth's shadow falls across the Moon, the edge of the shadow is curved—showing that the surface of the Earth must also be curved.

The records left by Thales, Aristotle and others were kept in a large library at Alexandria, in Egypt. Unfortunately this library was later destroyed—a loss which can never be made good. For many years the library was in the charge of Eratosthenes of Cyrene, a great scientist in his own right, who has earned his place in history as being the first man to measure the size of the Earth. From one of the books in the library he learned that at the time of the Summer Solstice—the 'longest day' in northern latitudes—the Sun was directly overhead as seen from the town of Syene (the modern Aswan), some way up the Nile, so that at noon the solar rays would shine directly into a well without casting a shadow. At this moment, however, the Sun was not overhead at Alexandria; it was 7 degrees away from the *zenith* or overhead point. A full circle contains 360 degrees, and 7 is about one-fiftieth of 360, so that if the Earth is a globe its circumference must be 50 times the distance from Alexandria to Syene. Eratosthenes measured this distance, and worked out that the distance right round the Earth must be about 24,850 miles. There is some uncertainty about this result, as his figures were calculated not in miles but in 'stadia', and the precise length of one stadion is not known; but his figure for the Earth's circumference seems to have been correct to within less than 100 miles.

The Greeks knew that the world is spherical, and they had an excellent idea of its size, but they found it difficult to believe that the Earth could be anything but the centre of the universe. This was a serious barrier to further progress. Even Aristotle was certain that the whole sky moves round the Earth. One or two astronomers —notably Aristarchus, who lived from about 310 to 230 B.C.—

ERATOSTHENES' METHOD OF MEA-
SURING THE SIZE OF THE EARTH. *At
noon at the time of the summer solstice, the Sun
was vertical at Syene, but not at Alexandria.
Eratosthenes measured the Sun's altitude at
this time, as seen from Alexandria, as 7
degrees away from the zenith, and was thus
able to measure the circumference of the Earth
with remarkable accuracy. It is significant
that the value which he gave was more correct
than that used many centuries later by
Christopher Columbus on his voyage of
discovery to the New World, which is a
remarkable tribute to Eratosthenes' theoretical
and practical skill*

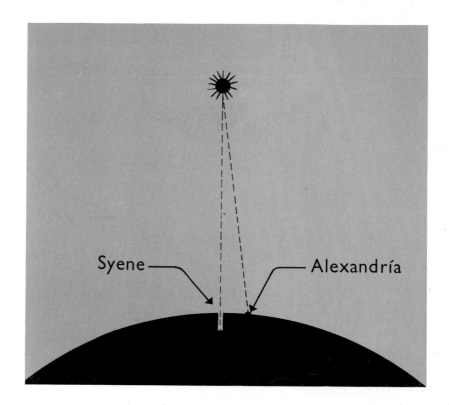

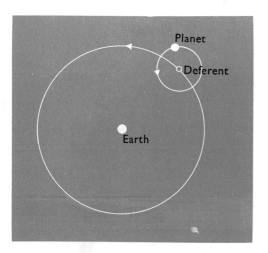

THE PTOLEMAIC THEORY. *According to
Ptolemy, a planet moved in a small circle or
epicycle, while the centre of this circle (the
deferent) itself moved round the Earth in a
perfect circle. This system was not invented
by Ptolemy, but its greatest development was
due to him. Ptolemy—who was an excellent
mathematician—realized that this comparatively simple arrangement would not explain
the actual movements of the planets in the sky,
and he was compelled to introduce extra
epicycles, thus making the whole system
clumsy and unwieldy. However, the Ptolemaic
theory was almost universally accepted by
scientists up to the time of Copernicus*

had the courage to suggest that the Earth revolves round the Sun,
but such ideas were highly unpopular, and the last two important
astronomers of ancient times, Hipparchus and Ptolemy, kept
firmly to the older theory.

Hipparchus lived about 150 B.C. Details of his life are unknown,
but it is obvious that he was a man of great ability, and he drew
up an important and remarkably accurate star catalogue. It was
Hipparchus, too, who discovered precession, the apparent movement of the celestial pole; and he worked out the distances of the
Sun and Moon more correctly than had been done before, though
the values which he gave were still much too small.

Claudius Ptolemæus, better known as Ptolemy, lived in
Alexandria. As with Hipparchus, nothing is known about his
career or personality, but science owes him a great debt. He produced a book in which he gave not only his own results, but also
those of the astronomers who had come before him, and it is this
book which has provided modern scholars with much of their
information about ancient science. It has reached us by way of
its Arab translation, and is generally referred to by its Arab name
of the *Almagest*, or 'greatest'.

Ptolemy was an excellent observer and mathematician, and he
undertook a thorough revision of Hipparchus' star catalogue. He
also investigated the movements of the bodies of the Solar System,
and concluded that the Earth lay in the centre, with the whole
heavens moving round it. The Moon was the nearest object in
the sky; then came the planets Mercury and Venus; then the Sun,
and then the more remote planets Mars, Jupiter and Saturn,
beyond which lay the stars. This arrangement is termed the
Ptolemaic System, although Ptolemy himself was not the first to
describe it.

THE UNIVERSE, ACCORDING TO PTOLEMY. *From an old print, 1600. The arrangement of the celestial bodies according to the Ptolemaic theory is clearly shown, though no attempt has been made to make the distances even approximately correct. However, Ptolemy realized that the actual motions of the planets must be complex; as we have seen, he was compelled to introduce numerous epicycles in an attempt to reach agreement with the observational data.*

'REVOLVING TABLE' FOR ESTIMATING THE POSITIONS OF THE ZODIACAL CONSTELLATIONS BETWEEN 7000 B.C. AND A.D. 7000. *Published in the famous book* Astronomicum Cæsareum *by P. Apian (Ingolstadt, 1540). The apparent movements of the stars are of course affected by the shifting of the celestial pole due to precession, so that conditions were not precisely the same in Apian's time as they had been in Ptolemy's. Yet even in 1540, the Ptolemaic theory of the universe was still generally accepted, and few scientists even considered questioning it*

The movements of the planets caused him a great deal of difficulty. It had always been supposed that the orbits of the heavenly bodies must be circular, since a circle was regarded as the 'perfect' form, and nothing short of perfection could be allowed in the sky. This was Ptolemy's view, but he realized that the observed motions of the planets could not be explained by the theory that they simply turned round the Earth in circular paths. Consequently Ptolemy worked out a system according to which each body moved in a small circle or *epicycle*, the centre of which itself moved round the Earth in a perfect circle. Even this would not suffice, and more and more epicycles were introduced, until the whole system became hopelessly clumsy and artificial.

Ptolemy never solved this problem, and after his death, about A.D. 180, the science of astronomy came almost to a standstill. Greece was no longer powerful, and the Roman Empire was crumbling; the Dark Ages came to Europe, and much of the old learning was forgotten. Many of the books in the Alexandrian Library were destroyed, and the remainder scattered. It is true that in Central America the people known as the Maya were making observations and working out an accurate calendar—but there was no communication between the Old and the New Worlds, and progress in European astronomy was halted.

Centuries passed. Then, slowly and painfully, astronomy was reborn—not for its own sake, but because of astrology. To make their predictions, the astrologers had to learn about the movements of the planets, and this knowledge could be gained only by careful observation. The Arabs took the lead, and some of their star catalogues were better than Ptolemy's. Then, as the peoples of Europe gathered into definite nations, the stage was set for the next phase in the history of astronomy.

4 the rebirth of astronomy

ARABIAN ASTROLABE. *This astrolabe, a typical example, was made in 1014 by Mustafa Ayyub*

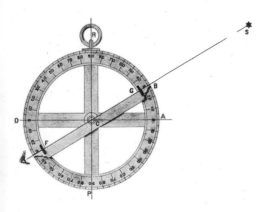

DIAGRAM TO SHOW THE USE OF AN ASTROLABE. *The axis DA is made level, and the star S is sighted along the direction FCBS; the observer's eye is shown behind F. The altitude of the star is then read off on the scale. In this case, the altitude amounts to 30 degrees*

THE ROMANS, WHO RULED much of the civilized world for so many centuries, did very little for astronomy. They did not share the Greek love of learning, and were concerned only with practical affairs.

One matter which did seem worth troubling about was the state of the calendar. The true 'year', or time taken for the Earth to go once round the Sun, is not exactly 365 days, but more nearly $365\frac{1}{4}$, so that to draw up a calendar which will not become out of step with the seasons is not as easy as it might appear. Julius Cæsar realized this, and instructed a Greek astronomer, Sosigenes, to form a more accurate calendar. Sosigenes did his work well; for instance he invented 'leap year', which took care of the extra quarter-day in the Earth's period of revolution. The 'Julian calendar' was not perfect, and has been further improved since, but it was quite good enough to satisfy the Romans.

In many ways it is unfortunate that scientific progress came to a halt for a long time after the death of Ptolemy about A.D. 180. It is even more unfortunate that so many of the old books were lost. However, some survived, and one of these was Ptolemy's *Almagest*, which reached Baghdad, capital of the caliphs, in the eighth century. It was translated into Arabic, along with various other volumes, and serious astronomy began once more.

The Arabs themselves proved to be skilful observers, and founded astronomical centres at Damascus and Baghdad. They set up scientific equipment for measuring the positions of the stars, and one Caliph, Al Mamon—son of Harun al-Rashid, of *Arabian Nights* fame—built a fine observatory. It was quite unlike a modern observatory, since telescopes still lay far in the future, but it contained an excellent library. By the time Al Mamon died, in A.D. 833, Baghdad had become the 'astronomical' capital of Europe.

Probably the most famous astronomer of the Baghdad school was Al-Battani, who was born about the middle of the ninth century. He was a particularly good mathematician, and made observations which compared favorably with those of Hipparchus and Ptolemy. Al-Battani also wrote an important book, the English title of which may be given as *The Movements of the Stars*.

Another skilful Arab astronomer was Al-Sûfi, who lived from 903 to 986. He, too, wrote a book—*Uranographia*, in which he dealt largely with the apparent brilliancies of the stars. This is an important matter, and is worth describing in a little more detail.

The stars are graded into classes or 'magnitudes' of apparent brightness. The scale may seem confusing, since the brightest stars have the smallest magnitudes; thus a star of magnitude 1 outshines a star of magnitude 2, and so on. The faintest stars visible to the naked eye, under ordinary conditions, are of magnitude 6. The scale may be compared with a golfer's handicap; the lower the handicap, the better the golfer.

Rigel in Orion is very brilliant, and on the modern scale its

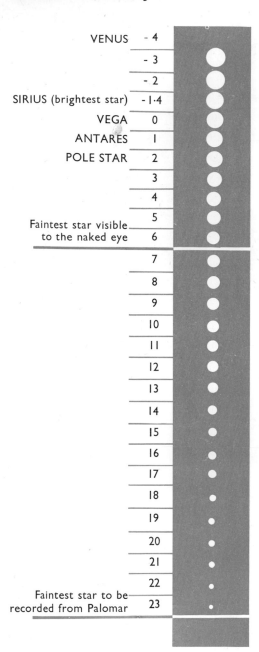

VENUS	– 4
	– 3
	– 2
SIRIUS (brightest star)	– 1·4
VEGA	0
ANTARES	1
POLE STAR	2
	3
	4
	5
Faintest star visible to the naked eye	6
	7
	8
	9
	10
	11
	12
	13
	14
	15
	16
	17
	18
	19
	20
	21
	22
Faintest star to be recorded from Palomar	23

STAR MAGNITUDES. *The magnitudes are shown here as disks, though the stars themselves appear only as points of light*

magnitude is reckoned as 0·1, or only one-tenth fainter than zero. The four brightest stars in the sky (Sirius, Canopus, Alpha Centauri and Arcturus) have negative magnitudes, that of Sirius being minus 1·4. On the same scale, the Sun has a magnitude of minus 27.

Nowadays it is possible to measure star magnitudes very accurately, and the world's largest telescopes can record objects of below magnitude plus 20. Al-Sûfi had to make his estimates simply by using his eyes, but most of his values agree well with those of today. On the other hand there are some interesting differences. He ranked Alhena, a star in the constellation of Gemini (the Twins) as of the third magnitude, but it is now brighter than the second, and there are other similar cases in which stars have apparently brightened up or faded away. It is difficult to be sure whether these changes are real, or are due to mistakes by Al-Sûfi and other observers of his time, but we do at least know that some of the stars are variable in brilliancy—and we must not be too quick to accuse Al-Sûfi of inaccuracy.

Many of the star names now in use are due to the Arabs. A typical example is Aldebaran, in the Bull, which means 'the following', since Aldebaran seems to follow the famous star-cluster of the Pleiades or Seven Sisters. Astronomical terms such as 'zenith' (the overhead point of the sky) are also Arabic in origin.

The Arabs did not confine themselves to measuring star positions and magnitudes. They carried out many other observations, and were particularly interested in eclipses. They also studied the movements of the Moon and planets. In this connection mention should be made of King Alphonso X of Castile, who called a number of Jewish and Arab astronomers to Toledo, and was responsible for the publication of the famous *Alphonsine Tables*, which contained data for planetary positions and the forecasting of eclipses, and which were used throughout Europe for the following 300 years.

The Mongol prince Ulugh Beigh, grandson of the Oriental conqueror Timur (more generally known as Tamerlane) also has his place in the history of astronomy. Ulugh Beigh founded a magnificent observatory at his capital of Samarkand, equipped with instruments which were the best of their day. Unfortunately for him, he was a firm believer in astrology, and this led to his death. He cast the horoscope of his eldest son, Abdallatif, and found to his alarm that the boy was destined to kill him. He therefore dismissed his son from Court and sent him into exile. Abdallatif had no wish to be set aside, and rebelled, finally murdering Ulugh Beigh and becoming king in his place.

The murder of Ulugh Beigh put an end to the Arab school of

ULUGH BEIGH MEMORIAL. *This memorial now stands in Samarkand, the site of Ulugh Beigh's old observatory*

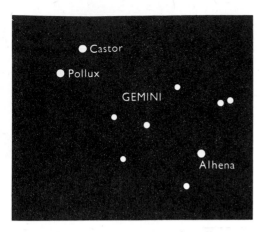

GEMINI, THE TWINS. *According to Al-Sûfi and others of his time, Castor used to be brighter than Pollux, but it is now half a magnitude fainter. Alhena, on the other hand, is now a magnitude brighter than as given by Al-Sûfi*

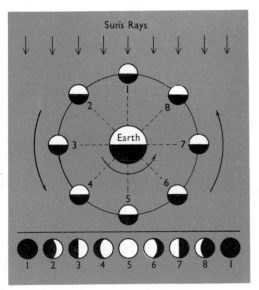

PHASES OF THE MOON. *The Moon has no light of its own, and depends on reflecting the rays of the Sun, so that half of it is luminous while the other half is dark. When the Moon has its dark face turned towards us, so that it is almost between the Earth and the Sun, we cannot see it at all, and this is what the astronomer terms 'New Moon'. (If the three bodies are exactly lined up, the result is an eclipse of the Sun, but owing to the tilt of the lunar orbit this does not happen at every New Moon.) In the diagram, New Moon is shown in position 1. At other times the Moon shows as a half (positions 3 and 7), three-quarter shape (4, 6), and full (5).*

astronomy, but by this time the old wish to learn had revived in Europe and observatories were set up in various places. The first was established at Nürnberg, in Germany, by Johann Müller, who is better known by his Latinized name of Regiomontanus. With his tutor, George von Peuerbach (Purbach) he revised the old Alphonsine Tables, and after Purbach's death he joined with his own pupil, Bernard Walther, to introduce new and better methods of observation. Printing had now been invented, and Regiomontanus set up his own press, so that he could publish astronomical information for the use of others. He continued to do so up to the time of his death in 1476.

Mention must also be made of 'the universal genius'—Leonardo da Vinci, who lived from 1452 to 1519. Leonardo was one of the greatest painters the world has known, and he was also a brilliant scientist. He was not a true astronomer, but he made one discovery which cleared up an old mystery. This concerned the Moon.

The cause of the Moon's phases, or apparent changes of shape each month from new to full, was not in the least mysterious. More puzzling was the fact that when the Moon shows as a crescent, the 'dark' portion can often be seen shining faintly, and giving the appearance known popularly as 'the Old Moon in the Young Moon's arms'. Leonardo realized that this must be due to light reflected from the Earth on to the Moon. Earth-shine is in fact nearly always seen whenever the crescent moon shines down from a clear, dark sky.

All this work, carried on in countries all over Europe, showed that astronomy had really 'woken up'. For instance star catalogues had been greatly improved, and careful observation had made it possible for astronomers to predict the positions of the planets for years in advance.

The main handicap to progress was the false theory that the Earth must lie in the centre of the universe, with all the other heavenly bodies moving round it in circular orbits. Astronomers had followed Ptolemy rather than Aristarchus, and so their ideas about the design of the universe were completely wrong.

A few scientists had their doubts. One was Nikolaus Krebs, the son of a German wine-grower, who was born in 1401 and died in 1464. His boyhood was unhappy; he ran away from home, and studied first at Heidelberg and then in Italy, becoming a well-known scholar and mathematician. He entered the Church, and became a Cardinal; he is often known as 'Nicholas of Cusa', since he was born at Cues on the Moselle. Krebs urged alterations in the calendar, and in a famous book called *De Docta Ignorantia* he suggested that after all, the Earth might not be lying at rest in the centre of the universe.

Not many astronomers of the time paid any attention to him, and for half a century after his death nothing more was heard of the theory of the 'moving Earth'. The next step—the realization that our own world is not, after all, an important body—was one which mankind found very hard to take.

Delhi Observatory

These photographs show one of the greatest observatories of pre-telescopic times, erected at Delhi in India. The right-hand picture shows part of the building, and a general view of the observatory is given below. An ancient observatory of this kind was, of course, very different from a modern astronomical observatory; all work had to be carried out with the naked eye only, and was therefore limited largely to positional measurements. Some of the measures made were, nevertheless, of surprising accuracy. The Delhi Observatory contained instruments which were capable of yielding very valuable results. Photographs by W. T. O'Dea

5 the design of the universe

DELHI OBSERVATORY. *An interior view.*
Photograph by W. T. O'Dea

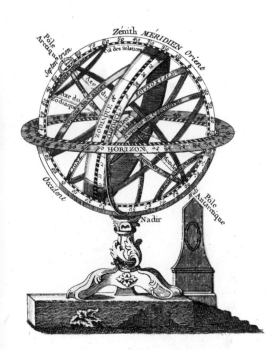

ARMILLARY SPHERE, *containing twenty circles; the Zodiac is clearly shown. From an old print*

SINCE GREEK TIMES, when scientists such as Ptolemy had rejected the idea that the Earth might move round the Sun, the development of astronomy had been held up. Aristarchus, who had hit upon the truth, had met with no support; neither did Nikolaus Krebs many centuries later. The man who altered all this, and finally changed our ideas about the universe, was a Polish cleric named Nicolaus Koppernigk, better known to us as Copernicus.

Copernicus was born at Thorn, on the River Vistula, in 1473. He studied at Cracow University, and then in Italy. Later he went back to his own country, and became Canon of Frauenberg in Ermland. In addition to his work for the Church he practised medicine, but his main interests were astronomical. Yet he was by no means a man who spent hour after hour looking at the stars. He was a theorist first and foremost, and he was concerned mainly with the design of the Solar System.

Early in his life he became very doubtful whether Ptolemy's system could be correct. The main trouble, as he saw it, was that the theory was so complicated. To account for the observed movements of the planets in the sky it had been necessary to add large numbers of small circles, or epicycles, until the scheme had become clumsy and artificial. In science, a simple and straightforward theory is generally more accurate than a cumbersome one, and Copernicus looked for some way of avoiding the complications which Ptolemy and his followers had been forced to introduce.

In one way Copernicus was better off than Ptolemy; he could make use of more accurate measures of the planetary movements, and he could be sure that these measures were not greatly in error. Finally he came to the conclusion that there was only one solution, the Earth must no longer be regarded as the centre of the universe. If it were assumed that the Earth, together with the other planets, moved round the Sun, then many of the complications would be removed at one stroke.

Today, we are so used to thinking of the Sun as the most important body in the Solar System that we find it hard to consider any other idea. Yet Copernicus knew that he was taking a bold step. He was saying that the astronomy then being taught in all schools and universities was utterly wrong. Worse still, he faced opposition from the Roman Catholic Church, which would certainly object to the idea that our world was not of supreme importance. Copernicus was himself a priest; and though he had worked out his theory by 1533, and written it down in a book which he called *De Revolutionibus Orbium Cælestium* ('Concerning the Revolutions of the Celestial Bodies'), he did not feel inclined to publish it.

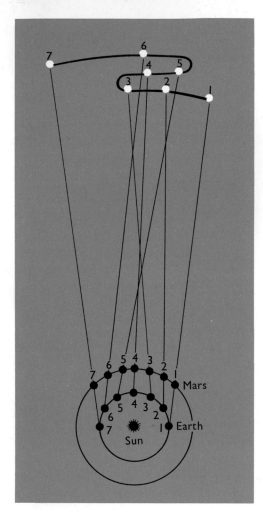

RETROGRADE MOVEMENT OF MARS. *The apparent path of Mars in the sky is given at the top of the diagram, and the actual relative positions of the Earth and Mars at the bottom. It will be seen that between positions 3 and 6 the Earth catches up Mars and passes it, so that for this period Mars seems to move in a retrograde or backward direction among the stars. Behavior of this sort was very difficult to explain on the old theory according to which the Sun moved round the Earth, and was one of the reasons why Ptolemy was forced to add further epicycles. Of the planets known in ancient times, Jupiter and Saturn behave in similar fashion, but the effects are less obvious because both these planets are so much farther away from the Sun, and their apparent movements in the sky are slower*

It is important here to consider the apparent movements of the planets, since this was the basis of all Copernicus' work. A planet does not move steadily among the stars in a straight line; it may sometimes seem to stand still for a few days, and then move 'backwards' for a short period before resuming its west-to-east journey. This apparent backward motion is known as *retrograding*.

Mercury and Venus have their own way of behaving, since they are closer to the Sun than we are—and even on Ptolemy's system they were assumed to be the closest bodies in the sky apart from the Moon. The remaining bright planets, Mars, Jupiter and Saturn, are more remote. Mars is shown in the diagram, but the same arguments apply to Jupiter and Saturn (as well as to Uranus, Neptune and Pluto, which were of course unknown in Copernicus' time).

For most of the time Mars seems to move from west to east among the stars, though the shift is so slow that it can be detected only over periods of hours. We now know that it is moving round the Sun in a larger orbit than that of the Earth, and also that it is travelling more slowly—only 15 miles per second, as against $18\frac{1}{2}$ miles per second for our own world. In the diagram we begin with the Earth at position 1 and Mars at position 1. By the time the Earth has moved to 2, Mars has reached 2, and so on. We can see that in positions 4 and 5, Mars is apparently moving across the sky in a retrograde direction, though in fact its real motion round the Sun is unaltered. In positions 6 and 7, the usual west-to-east movement has been resumed.

A planet which seems to perform a slow 'loop' in the sky was very hard to explain on Ptolemy's theory, and this was one of the reasons why Copernicus rejected the Earth-centred scheme.

Mars is well seen only at intervals separated, on an average, by 780 days (the interval is not quite constant). At such times the Earth is almost directly between Mars and the Sun. Mars therefore appears opposite to the Sun in the sky, and is said to be at *opposition*, as shown in the diagram on page 33.

Again we begin with the Earth at E1 and Mars at M1—the time of opposition. A year later, the Earth has completed one journey, and is back at E1; but Mars, moving more slowly and having farther to go, has not made a full circuit, and is in position M2. The Earth has to 'catch it up', and does so when the positions of the two bodies are E2 and M3 respectively, so that there is another opposition. The interval between successive oppositions is known as the planet's *synodic period*. When the planet is on the far side of the Sun, it is said to be in *conjunction*, and is above the horizon only during the hours of daylight.

It must not be supposed that Copernicus solved all the problems and drew up a really accurate plan of the Solar System. He certainly took the great step of placing the Sun in the centre, but he still believed that the orbits of the planets (including the Earth) must be circular; a circle was the 'perfect' form, and surely nothing short of perfection could be allowed in the heavens? This led to new difficulties, and Copernicus was forced to bring back epicycles to account for the observed motions of the planets. In fact, he was falling back into the trap which he had tried so

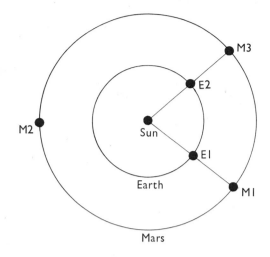

OPPOSITIONS OF MARS. *When the Earth is at E1 and Mars M1, the Sun, the Earth and Mars are in almost a straight line; Mars is opposite to the Sun in the sky, and is at* opposition. *A year later the Earth has returned to E1; but Mars, moving more slowly in a larger orbit, has not completed a full revolution, and lies at M2, so that it is unfavorably placed for observation. Another opposition does not occur until the Earth has 'caught Mars up', when the Earth will be at E2 and Mars at M3. This is why oppositions of Mars occur only at intervals of about 780 days (the synodic period of Mars)*

hard to avoid. He finally reached a scheme according to which the Solar System was made up of a central Sun and six planets moving round it in circular paths, with the Moon going round the Earth and with the fixed stars beyond the orbit of the most distant planet, Saturn.

Copernicus' book was ready, but remained unpublished, simply because he knew that it was certain to arouse violent criticism from the Church. Many people were aware of its existence, however, and tried hard to make its author give it to the world. Even the Archbishop of Capua, Cardinal von Schönberg, wanted it to be published. Georg Rhæticus, at one time Professor of Mathematics at Wittenberg in Germany, was also highly interested, and went to Frauenberg to hear about the new theory. He was forty years younger than Copernicus, and became his pupil, staying at Frauenberg for two years. It appears to have been largely because of Rhæticus' urging that Copernicus at last agreed to publish the complete book. Wisely he added a dedication to the Pope, Paul III, and Rhæticus took the manuscript to Nürnberg to have it printed.

In 1543 *De Revolutionibus* appeared, though the publisher, Osiander, had added an announcement to the effect that the Sun-centred theory was merely 'a mathematical fiction' which would be convenient for use in predicting the positions of the planets. Copernicus had not agreed to any such thing, but by now he was an old man, and seriously ill. In fact it is said, probably with truth, that the first printed copies of his great work reached him only a few hours before he died.

It is only too likely that had he lived, Copernicus would have found himself in serious trouble with the Church. This was what he had always feared, and his alarm was well-founded. At first only Rhæticus and his friend Erasmus Reinhold, also a Professor of Mathematics in Wittenberg University, dared to come out in open support of the new theory. The Roman Catholic authorities were strongly against it, though the real storm did not break until over half a century after Copernicus' death.

We must admit that Copernicus realized only part of the truth; he made many mistakes, and parts of his book were unsound. Yet he had found the essential clue, and for this reason alone he must be regarded as one of the greatest men in the history of astronomy.

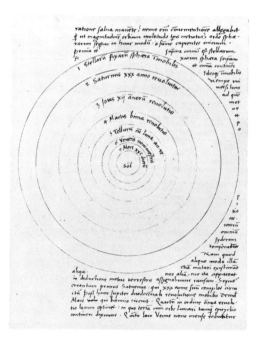

PAGE FROM COPERNICUS' GREAT BOOK. *A reprint of a typical page from the* De Revolutionibus Orbium Cœlestium *of Copernicus, published in 1543*

TYCHO BRAHE, *the great Danish observational astronomer*

TYCHO BRAHE'S QUADRANT. *With instruments of this kind, all built by himself, Tycho drew up his remarkably accurate star catalogue*

THREE YEARS AFTER the death of Copernicus and the publication of *De Revolutionibus*, a boy of very different character was born in Denmark. His name was Tycho Brahe, and in every way he was unlike Copernicus—except that he had the same love of astronomy, and the same urge to find out more about the celestial bodies.

Tycho's father, Otto Brahe, was of Swedish descent, though his family had lived in Denmark for some time. He was of noble birth, and held the post of Governor of Helsingborg Castle. His brother George was wealthy and childless, and Otto made a strange agreement that he would hand over his first-born son so that the boy could have an expensive education. When Tycho was born, Otto Brahe apparently regretted having agreed to give him up, and after a family dispute George took the baby away without permission. For a time the quarrels went on, but eventually they were patched up, and Tycho remained in the care of his uncle.

At the age of thirteen Tycho was sent to the University of Copenhagen. His uncle wanted him to become a statesman, but Tycho was not enthusiastic. Not long after reaching Copenhagen he watched a partial eclipse of the Sun, and this turned his attention to astronomy. He managed to obtain a copy of Ptolemy's *Almagest*, and within a year or so had managed to understand nearly all of it—even though he was still only a boy, and had not had much tuition in mathematics and science.

George Brahe was displeased, and sent Tycho away from Denmark, entering him at the German University of Leipzig. With him went a tutor, Vedel, who was given orders to see that Tycho gave up astronomy and returned to studies of statecraft. However, Tycho was not easily influenced, and he continued to work on astronomical theories. When he was seventeen he recalculated the apparent position of the planets in the sky, and found that the official tables of planetary movements were seriously wrong. At once he began making observations for himself. Vedel gave up trying to lead him back to statecraft, and about this time George Brahe died, so that Tycho was left to follow the career which he had chosen.

From Leipzig he went to the university at Rostock, and stayed there for some years. He was learning more about astronomy as he grew to manhood, but he was also convinced of the truth of astrology, and remained so to the end of his life. One curious story dates from 1566, when there occurred an eclipse of the Moon. Tycho watched it, and then announced that it foretold the death of the Sultan of Turkey, who was then the most powerful ruler in Eurasia. The Sultan did in fact die, and Tycho boasted of his astrological skill; but it then became known that the Sultan's death had occurred well before the eclipse of the Moon. It may appear strange that a man of Tycho's mental powers should believe in astrology, but sixteenth-century science was still riddled with superstition.

It also seems that Tycho was a hot-tempered, quarrelsome man. While at Rostock he had a dispute with another Danish nobleman, whose name has not been recorded, and the result was a duel, fought with swords in the middle of the night. The fight ended when Tycho had part of his nose cut off. He made himself a new part with gold, silver and wax, and apparently suffered no ill-effects.

So far Tycho's main work had not begun. The start of his true career may be said to date from 1572, when a new star appeared in the sky, and held the attention not only of Tycho but also of astronomers all over the world.

One of the most famous of all constellations is Cassiopeia, the 'Lady in the Chair'. It is easy to find, as the Great Bear and Polaris may be used as direction-finders to it; its five chief stars are arranged in the form of a W, and never set over Great Britain. It was here that Tycho's Star blazed up, on November 11, 1572, and it is worth recording the words of Tycho himself:

In the evening, after sunset, when, according to my habit, I was contemplating the stars in a clear sky, I noticed that a new and unusual star, surpassing the other stars in brightness, was shining almost directly above my head; and since I had, almost from boyhood, known all the stars of the heavens perfectly (there is no great difficulty in attaining that knowledge), it was quite evident to me that there had never before been any star in that place in the sky, even the smallest, to say nothing of a star so conspicuously bright as this.

Naturally enough he was filled with amazement. It had always been thought that the stars were unchanging, and for the moment he doubted the evidence of his own eyes. When others too saw the star, he knew that there could be no mistake. He began to make careful observations, and as the days passed by the star became even brighter, until it far outshone even the planet Venus. It was visible even during broad daylight. Then, slowly, it faded away. At last it fell below the sixth magnitude; and since telescopes had not then been invented, even the keen-eyed Tycho could follow it no further.

We now know what the star was. It was a *supernova*, a real stellar outburst, sending out as much luminosity as millions of Suns put together. Only three other supernovæ have been seen in our own star-system during recorded times—those of 1006 and 1054, observed by Chinese and Japanese astronomers, and (probably) the star of 1604.

A star, as we know, is a globe made up of intensely hot gas. Normally it shines steadily, and does not alter much over periods of thousands or even millions of years. In some cases, however, a star may suffer some tremendous internal disturbance, and will flare up suddenly, increasing its output of light and heat many thousandfold before dying back to its original brilliance. When a star behaves in this manner it is known as a *nova*, from the Latin word for 'new'. (The name is rather misleading, since a nova is not in fact a completely new star.) Normal novæ are not particularly uncommon. Several have been seen during the past few years; there was a bright one in 1934, and others in 1936, 1946, 1960, 1963, 1967 and 1968, while fainter novæ are fairly frequent.

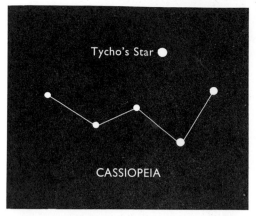

POSITION OF TYCHO'S STAR. *Tycho made careful observations of the brilliant supernova of 1572, which appeared in Cassiopeia not far from the famous 'W' of stars. This was the brightest supernova of the past thousand years. Only three others have been seen in our own Galaxy during this time; the 1054 star was certainly a supernova, and so was Kepler's Star of 1604, while it has recently been established that a star seen in 1006 in the constellation of Lupus (the Wolf) must also have been a supernova*

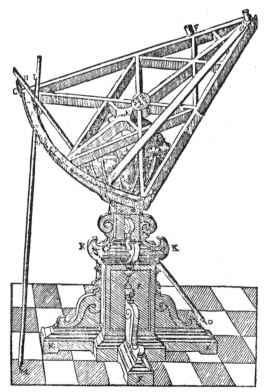

SEXTANT USED BY TYCHO BRAHE. *This sextant was in use by Tycho in 1577. It was one of the elaborate instruments which were set up on the island of Hven, where Tycho worked for so many years*

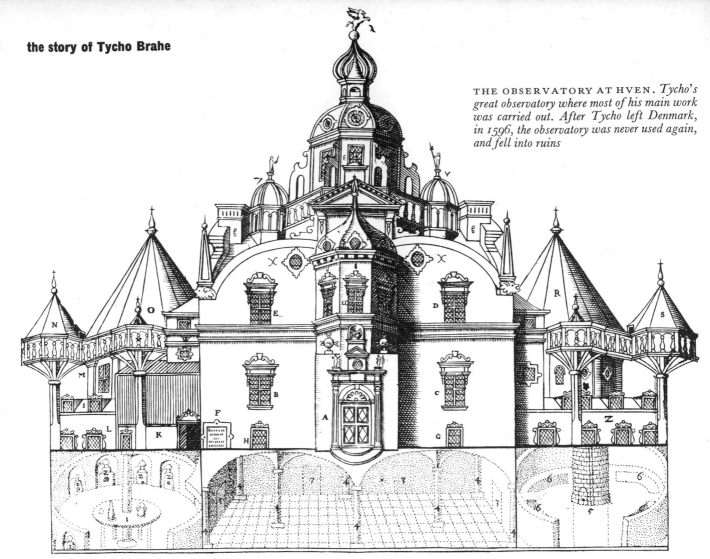

THE OBSERVATORY AT HVEN. *Tycho's great observatory where most of his main work was carried out. After Tycho left Denmark, in 1596, the observatory was never used again, and fell into ruins*

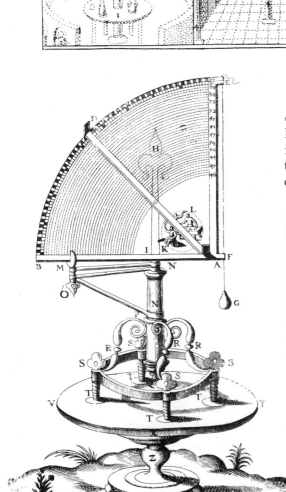

Tycho's Star was an outburst on a much grander scale. Evidently the explosion more or less destroyed the original star and hurled gaseous material into space in all directions. Today there is no visible trace of the original star itself, though our radio telescopes can pick up radiations which come from the remnants of the expelled gas.

Tycho knew very little about the nature of the stars, and he could not give a satisfactory reason for such an outburst, but at least the supernova made him determined to spend the rest of his life studying astronomy. He wrote a book about the star, *De Nova Stella*, and this book made him well known. By this time he was married, and he considered settling down in Basel, but then he received a generous offer from King Frederick II of Denmark. Frederick wanted Tycho to stay in Denmark, and granted him the little island of Hven in the Baltic, between Elsinore and Copenhagen, together with enough money to build an observatory and pay for its upkeep.

Tycho was quick to accept, and in 1576 he began the construction of his observatory—Uraniborg, the 'Castle of the Heavens'. It lay in the middle of a large square enclosure laid out as a garden, the corners of which pointed north, south, east and west. It contained a library and a chemical laboratory as well as

QUADRANT USED BY TYCHO BRAHE. *One of the great quadrants used by Tycho at Hven for his measures of star positions*

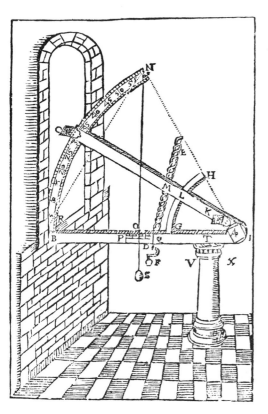

ALTITUDE INSTRUMENT USED BY TYCHO BRAHE. *This was yet another of the complex instruments made by Tycho and used at the observatory on Hven*

STJERNEBORG. *This is a recent view of Stjerneborg, the second of Tycho's two observatories on Hven. The buildings shown are of course modern, since nothing now remains of the original observatory. Photograph by Gösta Persson, 1958*

living apartments and the rooms for the instruments themselves. Later, in 1584, he built a second 'Castle of the Stars', Stjerneborg, in which some of the instruments were located below ground-level. The reason for this seems to have been that Tycho experienced trouble when the wind blew strongly and shook the instruments lying above ground. He also added a printing press and a paper-mill. Hven became a hive of scientific activity, and many distinguished people from all over the world visited it—among them James VI of Scotland, who afterwards became James I of England.

Tycho lived in magnificent style, and those who visited him were royally entertained. Banquets, games and hunts were held, and it is said that the guests were entertained by a dwarf whom Tycho kept specially for the purpose. On the other hand the islanders themselves were not well treated; Tycho was a harsh landlord. One of the less welcome buildings at Uraniborg was a prison in which he used to lock up those who would not pay their rents, or who displeased him in other ways.

The instruments themselves were by far the best of their time, and since Tycho was a most careful and accurate observer he obtained excellent results. He measured the positions of 777 stars, and drew up a catalogue; it is said that his star positions were never in error by more than 1 or 2 minutes of arc. When we remember that he had no telescopes, and that all his work had to be done with instruments without lenses, we can see how good an astronomer he must have been. He was still enthusiastic about astrology, however, and never began observing without dressing himself in special robes.

As well as drawing up his star catalogue Tycho measured the apparent movements of the planets, and it was these observations which proved to be so useful later on. He was also interested in the brilliant comet which appeared in 1577, and proved that it must be much more distant than the Moon. This was a step forward, since up to that time it was still thought possible that comets might be near at hand, and perhaps contained in the upper part of the Earth's atmosphere.

Tycho regarded Copernicus' theory that the Earth could revolve round the Sun as heretical. On the other hand he knew quite well that the movements of the planets could not be explained on Ptolemy's theory, and he suggested instead that the planets revolved round the Sun, while the Sun and Moon revolved round the Earth. This was not an entirely new idea, but it satisfied very few people apart from Tycho himself.

King Frederick of Denmark died in 1588, and in 1594 Tycho lost another of his supporters, the Danish Chancellor Kaas. Unfortunately he had made himself unpopular everywhere; the islanders of Hven hated him (as they had good reason to do), and the new Royal Court was much less friendly to him than the old one had been. Tycho would never see that he could be in the wrong, and eventually his supplies of money were cut off. In 1596, after having worked at Hven for twenty years, he left Denmark in anger, taking with him the more portable instruments from Uraniborg and Stjerneborg, and went to Germany. He never returned to his own country.

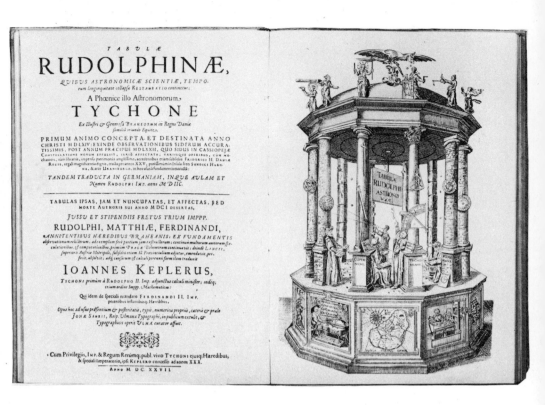

FRONTISPIECE OF THE RUDOLPHINE TABLES. *These tables represented Kepler's last astronomical work, and were published shortly before his death. Kepler's acknowledgement to Tycho is prominently featured on the title-page. The tables were so named in honor of Kepler's old benefactor, the Holy Roman Emperor Rudolph II, who was interested mainly in mysticism and astrology, but who also encouraged astronomical science. Rudolph himself had of course been dead for many years by the time that the tables appeared*

STATUE OF TYCHO BRAHE

For some time he had been in correspondence with Rudolph II, the Holy Roman Emperor, and Rudolph invited him to Bohemia, placing at his disposal the castle of Benatek, about twenty miles from Prague. Tycho accepted, and arrived there in 1599. At once he began setting up astronomical instruments, but it soon became clear that conditions at Benatek were very different from those at Hven. The Emperor himself was a curious character; he was mainly interested in astrology and 'alchemy' (the so-called making of gold out of baser elements), and he was both melancholy and incompetent. All through his reign his country was in a state of unrest, and finally, in 1611, he was deposed. Even at the time of Tycho's arrival the Emperor was short of money, and it was not easy for him to pay Tycho the salary which he had promised. Before long Benatek was given up, and Tycho settled in Prague itself.

By this time he had been joined by a younger man, Johannes Kepler, who was later to make the best possible use of the observations which Tycho had collected. The two men were not always on good terms, and it is very likely that the main fault was Tycho's, but they managed to work together for some time. Towards the end of 1601 Tycho became ill, and died on November 24, so that Kepler came into possession of his observations.

In every way Tycho Brahe was a picturesque figure. Hasty, intolerant, proud and often cruel, he was at the same time brilliantly clever, sincere and hard-working. Nothing now remains of his 'star castles' at Uraniborg and Stjerneborg, and indeed nobody else ever used them, so that after his departure from Hven they fell quickly into ruin. Yet it was Tycho's labors there which enabled his successor, Kepler, to prove once and for all that the Earth is an ordinary planet moving round the Sun.

7   legends in the sky

THE LEGEND OF PERSEUS AND ANDROMEDA. *This is one of the most famous of the mythological legends associated with the constellation patterns*

ARIES, THE RAM. *The first constellation of the Zodiac*

TYCHO BRAHE BECAME so famous that we tend to think that he was the only great astronomer of his day. Yet others too were working hard and skilfully. One of these, a German lawyer named Johann Bayer, produced a famous star atlas, the *Uranometria*, in which he lettered the stars according to their magnitudes, and produced a system of nomenclature which is still used.

As we have seen, the ancients divided the stars into groups or *constellations*, each named after a living creature, a common object, or else a mythological god or hero. Thus we have Ursa Major (the Great Bear), Sagitta (the Arrow), Orion, Hercules and many more. Ptolemy had listed a total of forty-eight constellations— twenty-one in the northern part of the sky and fifteen in the southern, as well as the twelve groups which make up the Zodiac. Not all the sky was covered; patches were left out, as well as the far southern regions which never rose above the horizon in Alexandria, where Ptolemy worked. Tycho added two more, one of which (Coma Berenices) is still to be found on our star maps, while Bayer was responsible for a dozen, all in the south. Still others have been added since.

At one time it was said that 'no astronomer seemed comfortable in his position' until he had named a constellation of his own. Not all were generally accepted—which is probably just as well. Johann Bode, a German astronomer who lived in the seventeenth century, listed nine new groups with long, clumsy names such as Sceptrum Brandenburgicum (The Sceptre of Brandenburg), Globus Ærostaticus (the Balloon) and even Officina Typographica (the Printing Press). All these have now been forgotten except by those who are interested in historical astronomy.

Some of the old legends about the constellations are charming, and the most famous of them all is that of the hero and the sea-monster.

According to the legend there was once a Queen, Cassiopeia, whose daughter, Andromeda, was exceptionally beautiful. Cassiopeia went so far as to boast that her daughter was lovelier than the sea-nymphs or Nereids, children of the powerful god Neptune. Neptune was enraged, and in revenge sent a monster to attack the Queen's country. The creature laid waste the shores, and before long Cassiopeia and her husband, King Cepheus, were in despair.

What was to be done? Cepheus consulted the Oracle, and was told that the only way to save his country was to chain Andromeda to a rock so that she could be eaten by the monster. There seemed no course but to agree. With heavy heart he gave the necessary orders, and Andromeda was left alone to await the coming of the terrible sea-beast.

It so happened that the hero Perseus had been on an expedition to kill the Gorgon, Medusa—a woman with snakes instead of hair, and whose glance would turn any living creature to stone. Perseus had been helped by the gods; he had been mounted upon

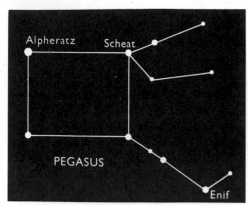

PEGASUS, THE FLYING HORSE.
Pegasus is a prominent constellation; it is marked by four stars arranged in a square. For some reason, however, one of the stars (Alpheratz) is now included in the neighboring constellation of Andromeda and is officially termed Alpha Andromedæ: it used to be known as Delta Pegasi. In mythology, Pegasus was the horse upon which the hero Bellerophon rode in his expedition against a terrible fire-breathing monster, the Chimæra

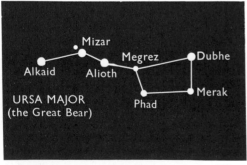

URSA MAJOR, THE GREAT BEAR. *The seven famous stars make up the 'Plough' or 'Big Dipper', but the whole of Ursa Major is a very extensive constellation*

a pair of winged sandals, and given a magic shield which would protect him from the Gorgon's stare. He had been successful in his quest, and had cut off Medusa's head. He was flying home, still carrying the head, when he saw Andromeda chained to the rock. At once he swooped down, and as soon as the monster appeared Perseus turned it to stone by confronting it with the Gorgon's head. He was rewarded with Andromeda's hand in marriage.

All the characters in the myth are to be found in the sky—Cepheus and Cassiopeia; Perseus, with the Gorgon's head marked by the famous 'winking star' Algol, which will be described later; even the sea-monster Cetus, which sprawls into the southern part of the heavens, and occupies a very large area of the sky.

This is only one of many such stories, but Bayer was concerned with a convenient nomenclature rather than legend. Many of the brightest stars, such as Sirius, Betelgeux and Rigel, had separate names, but it was clearly impossible to name each star in such a way, and something better was needed. What Bayer did was to take each constellation and allot each star in it a Greek letter, beginning with the first in the alphabet (Alpha) and ending with the last (Omega). Thus the brightest star in Andromeda became Alpha Andromedæ (Alpha of Andromeda); the second brightest should have been Beta; the third Gamma, and so on. Similarly, the brightest star in Ursa Minor, the Little Bear, became Alpha Ursæ Minoris, the official designation of the famous Pole Star. Actually, it sometimes happened that the stars became out of order. For example, Alpha Orionis, or Betelgeux, is not so brilliant as Beta Orionis, or Rigel. However, the system was clearly a good one, and Bayer's letters are still used.

We know very little about Bayer himself, and he accepted Tycho Brahe's star positions and magnitudes without question, but for his catalogue alone he deserves to be remembered.

By now the positions of the stars had been charted as accurately as was possible in pre-telescopic days, and their *right ascensions* and *declinations* were known with some precision. These terms too are still in use, and correspond more or less to longitude and latitude on the surface of the Earth.

We may begin by supposing the sky to be solid—the *celestial sphere*, as shown on page 41, with the Earth in the middle. Northward, the Earth's axis points to the North Celestial Pole, roughly marked by Polaris; there is no bright south polar star, the nearest naked-eye object being a fifth-magnitude star known as Sigma Octantis (Sigma in Octans, the constellation of the Octant).

On Earth, the latitude of a place is reckoned according to its angular distance north or south of the terrestrial equator. London, for instance, has a latitude of between 51° and 52°. We can imagine that the Earth's equator is projected on to the celestial sphere; this gives us the equator of the sky. The angular distance of a star north or south of this line is called the star's declination. Sirius has a declination of S. 16° 39', so that it is 16 degrees 39 minutes south of the celestial equator; Vega, the brilliant blue star in Lyra (the Lyre) is N. 38° 44', and so on.

If Polaris lay exactly at the pole of the sky its declination would

The Greek Alphabet

α	*Alpha*
β	*Beta*
γ	*Gamma*
δ	*Delta*
ε	*Epsilon*
ζ	*Zeta*
η	*Eta*
θ	*Theta*
ι	*Iota*
κ	*Kappa*
λ	*Lambda*
μ	*Mu*
ν	*Nu*
ξ	*Xi*
ο	*Omicron*
π	*Pi*
ϱ	*Rho*
σ	*Sigma*
τ	*Tau*
υ	*Upsilon*
φ	*Phi*
χ	*Chi*
ψ	*Psi*
ω	*Omega*

be N. 90°, and anyone standing at the North Pole of the Earth would see Polaris directly overhead. This is not quite the case; the declination is N. 89° 8′, so that the star lies 52 minutes of arc away from the true pole—a fact which navigators have to be careful to allow for in their calculations.

Declination, then, corresponds more or less to latitude on the

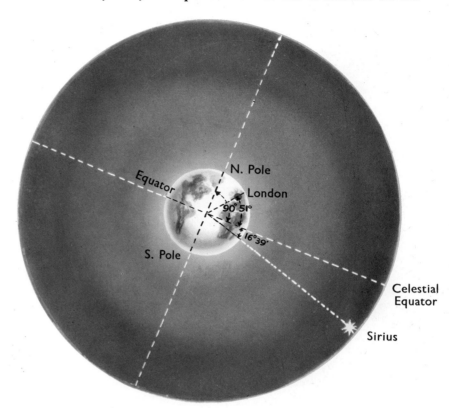

DECLINATIONS OF CELESTIAL OBJECTS. *The declination of a body in the sky is the angular distance of the body north or south of the celestial equator; the celestial equator is the projection of the Earth's equator on to the celestial sphere. Sirius, shown in the diagram, has a declination of south 16 degrees 39 minutes. The North Celestial Pole has, of course, a declination of N. 90 degrees; Polaris, the present Pole Star, lies within one degree of this. There is no conspicuous star within this distance of the South Celestial Pole. The declinations of the stars change very slowly owing to the effects of precession; the declinations of the Sun, Moon and planets naturally alter much more rapidly*

Earth, though—rather confusingly—the term 'celestial latitude' is used in another connection. Now let us turn to the celestial equivalent of longitude.

The Sun seems to travel along the Zodiac, completing one journey each year. Each March, about the 21st of the month, it crosses the celestial equator, moving from south to north. This moment marks the beginning of spring in the northern hemisphere, and the position where the Sun crosses the equator is known as the *Vernal Equinox*. It is also known as the First Point of Aries, since it used to lie in the constellation of Aries, the Ram. This is no longer true; as we have seen, the Earth's axis wobbles slightly and produces the effect known as precession, so that the Vernal Equinox shifts too. By now it has moved out of Aries altogether, and lies in the neighboring constellation of Pisces (the Fishes), though we still use the old name.

The Vernal Equinox is used as the starting-point for our measures of a star's right ascension. First we must refer to the *meridian* of any observing point on the Earth, which is a great circle in the sky passing through both celestial poles as well as the *zenith* or overhead point. The Vernal Equinox crosses the meridian once every 24 hours—often, of course, it does so in daylight—and is then

CULMINATION OF A STAR. A celestial body is said to culminate *when it reaches its greatest apparent height above the horizon*

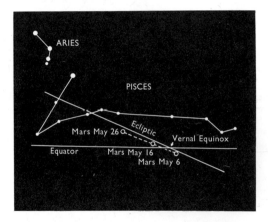

APPARENT PATH OF MARS IN MAY 1960. *Between May 6 and May 16, Mars passed by the First Point of Aries (Vernal Equinox) and also moved across the equator from the southern into the northern hemisphere of the sky*

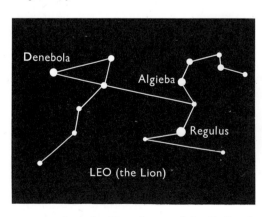

LEO. *Leo, the Lion, is one of the Zodiacal constellations. It contains one first-magnitude star (Regulus), and the 'Sickle', shaped rather like a question-mark twisted the wrong way round, is very easy to identify. Algieba and Denebola are of the second magnitude*

at its greatest height above the horizon, so that we say it *culminates*.

The difference in time between the culmination of the Vernal Equinox and the culmination of any particular star gives us the star's right ascension. Sirius, for instance, crosses the meridian 6 hours 43 minutes after the Vernal Equinox has done so, and we say that its right ascension or R.A. is 6h 43m. The right ascension of Vega is 18h 35m; of Rigel in Orion, 5h 12m.

It may sound confusing to measure the equivalent of 'sky longitude' in units of time instead of in angular measure, and various other systems are in use, but on the whole astronomers have found that this is the most convenient method. Once we know a star's right ascension and declination, we can fix its position in the sky just as accurately as we can fix a position on the Earth by quoting latitude and longitude.

The stars are so far away that they seem to keep in almost the same relative positions. The right ascensions and declinations change very slowly because of the effects of precession, but this is due to the wobbling of the Earth's axis, and has nothing to do with the stars themselves. We can also give the right ascensions and declinations of the Sun, Moon and planets, but obviously these will change rapidly. For instance, the positions for Mars during May 1960 were as follows:

May 6: R.A. 23h 47m, Decl. S. 2° 55′.
May 16:   0h 15m,   N. 0° 6′.
May 26:   0h 43m,   N. 3° 5′.

Between the 6th and 16th, then, Mars passed by the Vernal Equinox; it also crossed over the celestial equator from south to north.

Today we can work out the right ascension and declination of a star or planet to within a tiny fraction of a second of arc. Tycho, as we have seen, had to be content with an accuracy within 1 or 2 minutes of arc. Yet this is no reflection upon his skill. Considering that he had no telescopes, it is amazing that he was as correct as he was, and Bayer had every confidence in adopting the positions which he had given.

Kepler, too, had complete faith in Tycho—and it is ironical that he used the great Danish astronomer's work to prove the truth of the Copernican system which Tycho himself had so decisively rejected.

8   the laws of Johannes Kepler

JOHANNES KEPLER, *who drew up the famous Laws of Planetary Motion*

SUNDIAL, *constructed by Erasmus Habermeel, a mechanician at the Court of Kepler's benefactor the Emperor Rudolph II*

THE FIRST OF THE TRAGEDIES resulting from the 'Sun-centred' theory took place on February 17, 1600. Giordano Bruno, who was not strictly an astronomer but who nevertheless possessed deep scientific knowledge together with an inquiring mind, had been going around Europe teaching the truth of the Copernican theory that the Earth is a moving planet. This did not please the Church authorities, who insisted that the Earth must be the centre of all things and the most important body in the universe. Finally Bruno came to Rome, and was captured by the dreaded Inquisition. After being kept in prison for seven years, he was burned at the stake.

It would be wrong to suppose that Bruno was burned only because he had supported the idea that the Earth goes round the Sun; this was only one of his many crimes in the eyes of the Church. Yet it shows that to put forward such theories was extremely dangerous, and that the far-sighted scientist would be lucky if he did not find himself in serious trouble.

Tycho, of course, was in no danger at all, because up to the time of his death at the end of 1601 he still felt sure that the Sun moved round the Earth. He undoubtedly hoped that his pupil Johannes Kepler would use all the Hven observations to prove the clumsy and unlikely-looking 'Tychonic' theory. Kepler began his work with an open mind. Fortunately he was as brilliant as a theorist as Tycho had been as an observer.

Johannes Kepler was born at Weil der Stadt, in Württemberg (Germany) in 1571. His home life was the reverse of happy, and when he was four years old he became so seriously ill that he was left with a partially crippled hand, poor eyesight and a generally delicate constitution. It is therefore not surprising that he made very few astronomical observations himself. As he wrote, his eyes 'were not keen enough'.

He went to Tübingen University, planning to enter the Church. However, he soon turned his attention to astronomy, and in 1594 accepted a lecturing post at the University of Grätz. Two years later he published his first book, under a Latin title which may be translated as *The Forerunner of Dissertations on the Universe, Containing the Mystery of the Universe*. It contained a certain amount of useful science, though it also included a great deal of astrological superstition mixed in with some quite unsound ideas.

Tycho Brahe was among those who read Kepler's book, and he was highly impressed by it. Here, surely, was a man who would be able to make proper use of the Hven observations? Tycho invited Kepler to join him and become his assistant; Kepler was glad to accept, since there were religious troubles at Grätz, and it was impossible for him to stay there. In 1600 the young German mathematician joined the old Dane at Prague. Less than two years later Tycho died, and Kepler succeeded him as Imperial Mathematician to Rudolph II.

Now Kepler began his main life's work. Using the Hven

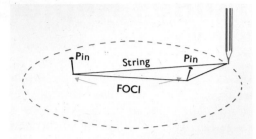

DRAWING AN ELLIPSE. *An ellipse may be drawn quite simply. Fix two pins into a piece of board, and join them with a thread, leaving a certain amount of slack. Then put a pencil through the thread, and trace out a curve. This will be an ellipse, and the pins will mark the* foci. *The wider apart you put the pins, the more* eccentric *will be the ellipse. If the two pins come together at a single point, the eccentricity is zero, and the ellipse becomes a circle. The orbits of the main planets known in ancient times are almost circular, but not quite—and it was this slight departure from circularity which enabled Kepler to prove the truth of the heliocentric theory as opposed to the Ptolemaic*

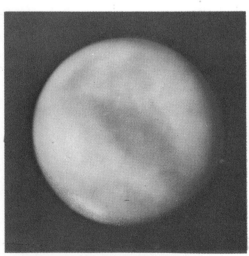

MARS, *photographed with the 60-inch reflector at Mount Wilson. South is at the top. The Syrtis Major is visible to the left*

observations, he had to decide whether Copernicus had been right in saying that the Earth moved round the Sun, or whether there could be some other explanation.

Tycho's star catalogue was by far the best yet made, and demonstrated his superlative skill as an observer. The relative or *proper* motions of the stars are so slight that for most purposes we may neglect them—and Kepler, indeed, did not know that such proper motions existed at all. He based his work on the apparent wanderings of the planets, particularly Mars. Fortunately, Tycho had made very accurate measurements of the positions of Mars, and Kepler was quite sure that he was justified in placing complete trust in them.

His task proved to be even more difficult than he had expected. The movements of Mars and the other planets could not be accounted for on the theory that they moved round the Earth; neither did the positions fit the idea that the planets, and the Earth, moved round the Sun in circular orbits. Kepler tried hard to make Tycho's observations fit, but for many years he failed to do so. The measured positions of Mars almost agreed with theory —but not quite. And then, at last, Kepler hit upon the truth. The planets move round the Sun indeed, but they do so in orbits which are not circles, but ellipses.

Kepler found that the planetary orbits were ellipses of very slight eccentricity—in fact, they were almost circular, but the slight difference between circle and ellipse allowed Tycho's observations to fall beautifully into place. To show what is meant, let us give some of the modern measures of planetary distances from the Sun. This means that the distance between the Earth and the Sun varies by about $1\frac{1}{2}$ million miles from its average value of 93 million miles.

DISTANCES OF PLANETS FROM THE SUN

Venus: maximum 67,600,000 miles; mean, 67,200,000 miles; minimum, 66,700,000 miles.

Earth: maximum 94,600,000 miles; mean, 93,000,000 miles; minimum, 91,400,000 miles.

Mars: maximum 154,500,000 miles; mean, 141,500,000 miles; minimum, 128,500,000 miles.

(*These figures are given in round numbers. The mean distance between the Earth and the Sun, known as the astronomical unit, has recently been remeasured by radar and radio astronomy methods; the latest value is approximately 92,956,000 miles.*)

Oddly enough, we are at our closest to the Sun in December, when the northern hemisphere is experiencing winter. The diagram will explain why this is so. In December, the Earth's North Pole is tilted away from the Sun, and so the solar rays fall at an angle; in July, the north part of the Earth is tilted towards the Sun, and the rays are more direct. Our seasons are due mainly to the inclination of the axis, and not to the Earth's changing distance from the Sun.

Venus has an orbit of even lower eccentricity, but Mars moves in a less circular path. This was fortunate for Kepler, since Mars was the world upon which he concentrated his main attention.

Now that he had taken the first real step, Kepler was able to draw up his three famous Laws of Planetary Motion. The first two were published in 1609, and the third nine years later. All three are so important in astronomical theory that they are worth describing in rather more detail.

Law 1 we have already met; it states that the planets move round the Sun in ellipses, the centre of the Sun being placed in one focus of the ellipse, while the other focus is empty.

Law 2 states that the *radius vector* of the planet sweeps out equal areas in equal times. The radius vector is an imaginary line joining the centre of the Sun to the centre of the planet. This is illustrated in the diagram, in which the Sun is represented by S. Suppose that the planet moves from A to B in the same time that it takes to move from C to D. Then the blue area ASB must be equal to the yellow area CSD. In other words, a planet moves quickest when it is closest to the Sun. This naturally applies to the Earth; in December we are travelling slightly faster than in July. (To make the diagram clear the orbit has had to be drawn as rather eccentric, whereas—as we have seen—the real orbits of the planets are very nearly circular.)

Law 3 states that for any planet, the square of the *sidereal period* (time taken to complete one journey round the Sun) is proportional to the cube of the planet's mean distance from the Sun. This sounds rather confusing, but it means that there is a definite relationship between the distance of a planet and the time which it takes to go once round the Sun.

This important third Law was published in a book called *Harmonices Mundi* ('The Harmonies of the World'), which, as usual, contained valuable scientific discoveries mixed up with astrological and mystical nonsense. It should be understood that Kepler, as Imperial Mathematician, had to be an astrologer too—officially at least. His benefactor, the Emperor Rudolph II, was more interested in astrology than in true astronomy.

Unfortunately Kepler was finding things increasingly difficult. Rudolph II was forced to abdicate, and died not long afterwards. Money was very short, and Kepler did not often receive his promised salary, so that in 1612 he moved to Linz, in Austria, and began lecturing at the University there. About this time his wife died, and also one of his three children. Kepler later married again.

There were further troubles in 1620, when his mother, Catherine Kepler, was arrested on a charge of witchcraft. In view of the period in which she lived, this was hardly surprising; she seems to have been a most unpleasant woman, and may well have looked rather like the popular idea of a witch. Johannes Kepler fought energetically to secure her acquittal, and finally he succeeded, though his mother did not live for long after her release from prison.

He was carrying on with his scientific work, and it would be wrong to suppose that his fame rests only upon his Laws of Planetary Motion. He wrote an important book about comets, and another in which he gave a full outline of the Copernican system. Strangely, perhaps, the Church authorities did not trouble him on this account, but in 1626 religious difficulties led to his being forced to leave Linz. Meanwhile he was working upon improved tables for planetary positions, and he called them the Rudolphine Tables

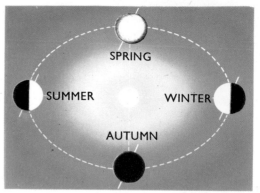

THE SEASONS. *This drawing relates to the northern hemisphere of the Earth; in the southern hemisphere conditions are of course reversed. During northern summer, the northern hemisphere of the Earth is tilted toward the Sun, and the rays strike this hemisphere more directly. At this time the Earth is actually at its greatest distance from the Sun (over 94,000,000 miles)*

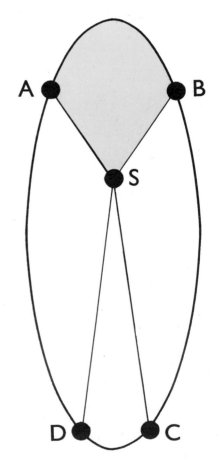

KEPLER'S SECOND LAW. *S represents the Sun; A, B, C and D a planet in four different positions in its orbit. According to Kepler's Law, a planet moves at its quickest when at its closest to the Sun*

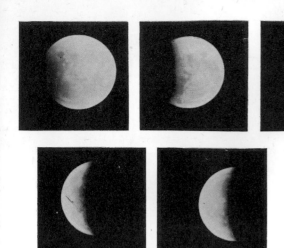

PHOTOGRAPHS OF AN ECLIPSE OF
THE MOON. *Totality occurred between the
fifth and sixth photographs. Generally the
Moon is plainly visible even when totally
eclipsed, though it is on record that at some
eclipses the Moon has disappeared completely
for a short period*

in honor of his old benefactor the Emperor Rudolph II. But he
was still desperately short of money, and it was almost impossible
for him to obtain the salary due to him as Imperial Mathematician.
At last, in 1630, he set out on a journey to press his claim to the
amounts due to him; but travelling was too much for him in his
weakened state, and on November 15 he died.

Though Kepler was not a true observer himself, mainly because
of his poor eyesight, he was very interested in the new telescopes
invented in 1608 and first turned to the heavens less than two
years later by Galileo. He also made some other notable advances
in astronomical theory. For instance, he explained why the Moon
seems to turn a coppery or reddish color during a lunar eclipse.

As we have seen, a solar eclipse is caused when the Moon passes
in front of the Sun and blots out the brilliant solar disk. An eclipse
of the Moon is quite different in nature, as is shown below.
The Earth, like any other solid body, casts a shadow. When
the Moon moves into this shadow its source of direct sunlight is
cut off; the Sun, Earth and Moon are then lined up, with the
Earth in the middle position. Usually, however, the Moon does
not disappear completely. The Earth's blanket of atmosphere
bends or *refracts* some of the sunlight on to the lunar surface, and
instead of vanishing the Moon becomes dim and, generally,
reddish-colored.

Lunar eclipses may be either total or partial. They are not
regarded as of much importance astronomically, but they are
beautiful to watch, and at any one place on the Earth they are
seen more frequently than eclipses of the Sun. This is because a
lunar eclipse, when it occurs at all, may be seen from a complete
hemisphere of the Earth. Binoculars or a small telescope will show
the coloration well, and it is worth remembering that this coloration was first explained by Johannes Kepler more than three
hundred years ago.

THEORY OF AN ECLIPSE OF THE
MOON. *An eclipse occurs when the Moon
passes into the shadow cast by the Earth*
[*This diagram is not to scale*]

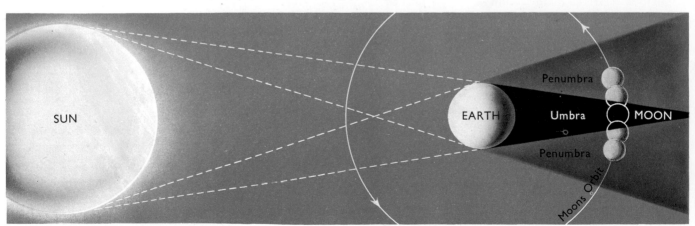

GALILEO. *A portrait of Galileo Galilei, the great Italian scientist who was the first great astronomer to use a telescope. Galileo was also a pioneer of experimental mechanics*

THE LEANING TOWER OF PISA. *It has been said Galileo dropped stones off the top of the famous Leaning Tower of Pisa, though as a matter of fact he never carried out this experiment. Photograph by Dominic Fidler*

THE YEAR 1609 MUST be regarded as one of the most important in the whole story of astronomy. It saw the publication of Kepler's first two Laws, based on the observations made long before by Tycho Brahe. More important, however, was the first use of the telescope in studying the celestial bodies.

Up to this time astronomers had to depend solely upon their eyes. They could measure the apparent positions of the stars and planets, but they could see them as nothing more than points of light. The Chinese and other ancient peoples had reported occasional dark patches on the face of the Sun, and the Moon's disk was clearly mottled with bright and grey areas, but it was impossible to find out anything useful concerning the surface features. All this was changed in a period of only a few months.

The principle of the telescope was first discovered by a Dutch spectacle-maker, Hans Lippershey. He was not an astronomer, and probably he never thought of using the new instrument to look at the heavens; but before long, news of his discovery spread. In particular it came to the ears of Galileo Galilei, Professor of Mathematics at the Italian University of Padua.

Galileo was born in 1564 (the same year as Shakespeare), and at the age of seventeen was sent to Pisa University to study medicine. He soon gave up this idea; he was much more interested in mathematics and experimental science, and when he was only eighteen he made his first important discovery. He found that when a pendulum is set swinging, the time taken for one complete to-and-fro movement depends entirely upon the length of the support and upon other characteristics of the pendulum itself. Moreover, the time taken for a full oscillation does not alter even when the pendulum has nearly stopped. A pendulum therefore is extremely valuable as a means of measuring intervals of time. All through his life Galileo tried to build a true pendulum clock, and actually designed one, though it is unlikely that he ever built it.

In 1589 he was appointed Professor of Mathematics at Pisa, and his fame grew quickly, not only because of his discoveries but because he was so clearly unwilling to take anything 'on trust'. For instance Aristotle, the Greek scientist who had lived nearly 2,000 years before, had said that a heavy body must fall to the ground more rapidly than a lighter one, and nobody had ever questioned this until Galileo did so. It is often said that he climbed the famous Leaning Tower of Pisa and dropped stones of different weights from the top, showing that they hit the ground at the same moment. Actually, this story (like so many of its kind) is not true, but Galileo would have been quite capable of carrying out the experiment if he had thought it worth while. A stone will certainly drop more rapidly than a feather, but this is because of the resistance caused by the air. Were there no atmosphere round the Earth, the stone and the feather would fall at exactly the same rate.

TWO OF GALILEO'S TELESCOPES, *preserved in the Tribuna di Galileo in Florence. A broken object-glass, with which the four satellites of Jupiter were discovered, is mounted in the centre of the ivory frame*

THE PLEIADES, *photographed in red light with the 18-inch Schmidt telescope at Palomar*

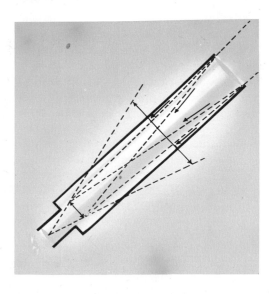

In 1592 Galileo became Professor of Mathematics at Padua, and went on with his experimental work. By this time he had apparently become sure that Copernicus had been right in saying that the Earth moves round the Sun, but as yet he had made no astronomical discoveries himself.

The appearance of a brilliant 'new star' in 1604 reawakened Galileo's interest in the skies. Like Tycho's Star of 1572 it may well have been a supernova, and it became so bright that nobody could overlook it. (It is often known as Kepler's Star, because Kepler was among those who observed it.) Galileo made careful measurements of its position, and satisfied himself that it was much more remote than any of the planets, though he could not be expected to understand its true nature.

Then, while still at Padua, he received the news of Lippershey's discovery. Let us quote Galileo's own words, written later in his famous book *Sidereus Nuncius* ('The Sidereal Messenger'):

About ten months ago a report reached my ears that a Dutchman had constructed a telescope, by the aid of which visible objects, although at a great distance from the eye of the observer, were seen distinctly as if near . . . At length, by sparing neither labor nor expense, I succeeded in constructing for myself an instrument so superior that objects seen through it appear magnified nearly a thousand times, and more than thirty times nearer than if viewed by the natural powers of sight alone.

Telescopes of this kind are known as *refractors*, and the principle of a modern refractor is shown in the diagram.

The light from the Moon, or whatever object is to be studied, falls upon a glass lens known as an *object-glass*. This bends the light-rays and brings them together at a point known as the *focus*, where an image of the Moon is formed. This image is then magnified by a second lens known as an *eyepiece*. The distance between the object-glass and the focus is known as the *focal length*.

Galileo's first instrument was feeble judged by modern standards, and had an object-glass only 1 inch in diameter, whereas the largest refractor at present in use, that at Yerkes Observatory near Chicago, has an aperture of 40 inches. Yet Galileo was able to make a whole series of amazing discoveries, and it was his work during the years from 1609 to 1619 which marks the beginning of telescopic astronomy.

It has been suggested that Galileo was not the first man to use a telescope in this way. It is just possible, though unlikely, that an Englishman named Leonard Digges—about whom we know almost nothing apart from the fact that he died in 1571—made some sort of optical instrument, and similar claims have been put forward for Roger Bacon and for an Italian named Porta. Moreover there were two other men who certainly began telescopic work at about the same time as Galileo, and independently of him. One was an Englishman, Thomas Harriot, who once acted as tutor to Sir Walter Ralegh; the other a German, Simon Mayer (better known as Marius). But however this may be, Galileo followed up his observations with such skill and energy that he is certainly entitled to the main credit.

PRINCIPLE OF THE REFRACTOR. *The light is collected by the object-glass*

Naturally enough, the first thing that he looked at was the Moon. At once he saw that the lunar surface is rough and mountainous. Along the *terminator*, or boundary between the daylight and night hemispheres, he saw shining points, and realized that these could be due only to mountain-tops catching the sunlight, so giving the terminator a rough and uneven appearance. He saw the grey plains, which later became known as 'seas', though Galileo himself did not believe that there was any water in them; again he has been shown to be correct. He described the valleys, and also saw the great circular formations which we now call craters. He drew up a telescopic map of the Moon's surface, which was the first apart from Harriot's; and he tried to measure the heights of the peaks, though his results were not accurate.

The next thing he found was that the stars did not show obvious disks in his telescope. No matter what the magnification, they still appeared as tiny points of light. Of course they seemed more brilliant, and more of them were visible; he could go down well beyond the sixth magnitude. He gave, as a typical instance, the case of the famous star-cluster known as the Pleiades, near Aldebaran in the Bull. To the naked eye six stars are easily visible on a clear night, but Galileo's telescope showed him at least forty, and we now know that the cluster includes over two hundred and fifty faint stars.

It is not surprising that Galileo failed to see stellar disks. Even with our great modern telescopes a star still appears as a tiny point. This is because all the stars are so far away. They are not really small, but if you use a telescope and see a star as a balloon of light you may be sure that either your telescope is not properly focused or that there is something else wrong with it.

Galileo then turned his attention to the Milky Way, the glorious band of radiance which stretches across the sky and is a lovely spectacle on a dark night. He found that it was made up of 'a mass of stars', apparently crowded close together. Again he was correct, as any modern binoculars will show.

In 1610 Galileo first looked at the planets through his telescope, and again he made some remarkable discoveries. On January 7 he found that Jupiter was attended by 'three little stars'. Within a few weeks he had established that there were four such objects, and that they were not ordinary stars; they were moons or *satellites*, revolving round Jupiter just as our Moon revolves round the Earth.

This discovery seems to have excited Galileo more than any other. The system of Jupiter seemed to be a Solar System in miniature, and he became more than ever convinced that Copernicus had been right. He found a still more positive proof when he saw that the planet Venus shows *phases*, or changes of shape, similar to those of the Moon. Sometimes Venus was a crescent, sometimes a half, and sometimes almost a full disk.

Let us look back for a moment at the old Ptolemaic system. On this theory, both Venus and the Sun moved round the Earth, but to explain the apparent movements of Venus it had been necessary to suppose that the planet had a complex motion. It moved round a point or *deferent*, lettered D in the diagram, while D itself moved round the Earth in a perfect circle. If E

VENUS AT CRESCENT PHASE. *From an observation by Patrick Moore, made on September 30, 1959, with an 8½-inch reflector, × 300. Drawing by D. A. Hardy*

A MODERN REFRACTOR. *This is the refractor at the Sternberg Institute in Moscow; the object-glass is 8 inches across. Photograph by Patrick Moore, 1960*

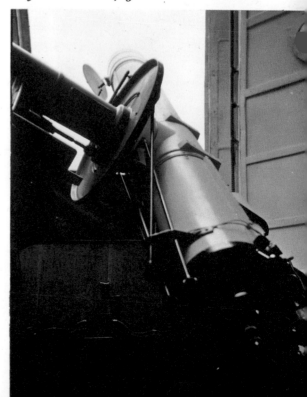

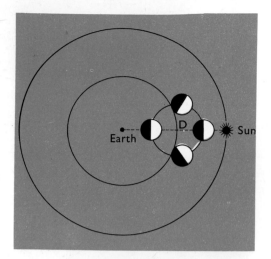

MOVEMENTS OF VENUS ACCORDING
TO PTOLEMY'S THEORY. *According to
Ptolemy, both Venus and the Sun move round
the Earth, and the line joining the Earth, the
Sun, and the deferent of Venus is always
straight. This would mean that Venus could
never show as a half or full disk; it would
always be a crescent when visible at all.
Though there have been many reported cases of
the crescent shape of Venus being visible to the
naked eye, it is certain that nobody could
follow the changing phases without optical aid,
and until the invention of the telescope there
could therefore be no certain proof. When
Galileo was able to use a telescope to study
Venus, he saw at once that the old theories
must be wrong, since they could not possibly
account for Venus' behavior*

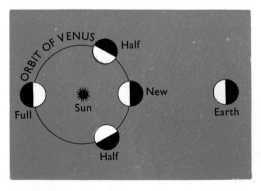

CORRECT THEORY OF THE MOVE-
MENTS OF VENUS. *The planet may be
seen as a crescent, half or full disk*

[This drawing is not to scale]

represented the Earth and S the Sun, then according to Ptolemy,
the line EDS was always straight.

We know that Venus shines by reflected sunlight, so that only
half the planet is luminous; the other half—the half turned away
from the Sun—is dark. On Ptolemy's theory, then, it is clear that
Venus could never show as a full disk, or even a half. Yet it does—
and Galileo saw why.

The lower diagram shows the true position. Both Venus and the
Earth go round the Sun, the Earth in $365\frac{1}{4}$ days and Venus in $224\frac{3}{4}$
days. When Venus is in the position marked 'New', it is almost
between us and the Sun, and cannot be seen. At *dichotomy* (a Greek
term meaning 'cut in half') it appears as a half disk, and at full it
shows a circular disk, though it is then on the far side of the Sun
and cannot be seen without a telescope.

This was decisive. Ptolemy's theory was wrong; therefore, the
Sun-centred idea must be right. Galileo was never inclined to
keep his opinions to himself, and he openly taught the truth of the
Copernican doctrine. The publication of *Sidereus Nuncius*, not long
afterwards, made him world famous. It also led to the first signs
of trouble with the Catholic Church.

Many people were doubtful whether Galileo could be right in
all he said; he was even accused of bewitching his telescopes.
Moreover, there were some things which were hard to explain.
The planet Saturn, in particular, showed a triple appearance
which later vanished, and it was not until 1656 that the Dutch
astronomer Huygens accounted for this satisfactorily. Galileo also
observed sunspots, showing that the Sun was by no means the
unblemished globe which it had been thought to be. But the real
trouble was Galileo's outspoken support of Copernicus. In 1615
the Church authorities stated 'that the doctrine that the Sun was
the centre of the world and immovable was false and absurd,
formally heretical and contrary to Scripture', and Galileo was
officially warned to alter his views. Meanwhile he had resigned
his post at Padua, and had settled in Florence as mathematician
to the Grand Duke of Tuscany. For a while he was left more or
less in peace.

Then, in 1632, he published his great book *Dialogue Concerning
the Two Chief World Systems—Ptolemaic and Copernican*. This is in the
form of a conversation between two imaginary philosophers, and
Galileo showed his faith in Copernicus so clearly that the Church
took action. In 1633 he was summoned to Rome, and forced to
'abjure, curse and detest' his supposedly false view that the Earth
moves round the Sun. He was too wise to protest—at least openly;
in those days the Church was all-powerful, and it was only thirty-
three years earlier that Giordano Bruno had been burned at the
stake.

By now Galileo was an old man, and the rest of his life was
passed at his home in Arcetri. He was carefully watched by the
Church authorities, and expressly forbidden to carry on with his
astronomical work. Moreover his eyes were failing, and during his
last years he became totally blind. Even so he still managed to
accomplish great things, and one major discovery which he made
during this period concerns the rotation of the Moon.

The Moon moves round the Earth in a period of twenty-seven

THE MILKY WAY IN CYGNUS. *Photographed at Lowell Observatory with a 5-inch lens, exposure 3 hours, on April 13, 1930*

THE LUNAR MARE CRISIUM. *The Mare Crisium, or Sea of Crises, is one of the smaller but most conspicuous of the lunar maria. It lies fairly close to the limb, and its appearance is therefore markedly affected by libration. Lick Observatory photograph*

MAKING A SIMPLE TELESCOPE. *It is simple enough to make a small refractor out of spectacle-lenses and cardboard tubes. The cost is low, and the construction takes only an hour or two*

and one-third days. It also spins once on its axis in twenty-seven and one-third days, so that it keeps the same face towards us all the time; the effect is much the same as for a man who walks round a chair, turning so as to keep his face towards the chair—in which case the back of his neck will never be turned 'chairward'. From the Earth, part of the Moon is permanently invisible, and remained unknown until the Russian rocket Lunik III passed round the Moon, in 1959, and sent back photographs of the hidden side.

Yet as Galileo found, the Moon seems to sway very slowly to and fro in the sky. Sometimes a little of one side is revealed; sometimes a little of the opposite side. The reason is that while the Moon spins on its axis at a constant rate, its speed in its path round the Earth changes. It moves in an orbit which is not a circle, but an ellipse, and naturally it moves fastest when it is closest to us. (The actual distance from the Earth varies between 221,460 and 252,700 miles.) Therefore the position in orbit becomes periodically 'out of step' with the amount of axial spin, and we can see for some distance round alternate edges. From the Earth we can examine a total of 59 per cent of the lunar surface, though of course we cannot see more than 50 per cent at any one time. This effect is known as the Moon's *libration in longitude*.

Galileo died in 1642, at his home in Arcetri. By that time he was worn-out and sightless, but he had accomplished great feats, and had ensured that his name will never be forgotten.

It is possible to make a telescope not unlike Galileo's, at very low cost. Go to an optician and buy a spectacle-lens from 1 to 2 inches in diameter and about 2 feet focal length. Ask also for a smaller lens of shorter focal length, to act as the eyepiece. Fix them into cardboard tubes, as shown here, using glue and slots; make some sort of rough mounting, and you will have an instrument powerful enough to show the mountains of the Moon, the stars of the Milky Way, and the satellites of Jupiter—almost as well as Galileo himself saw them so long ago.

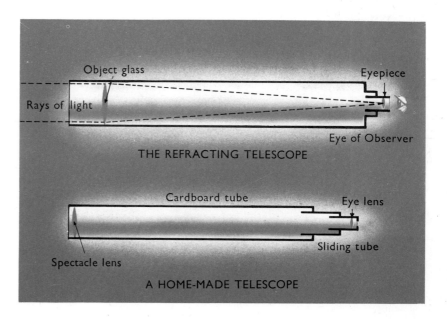

Object glass
Eyepiece
Rays of light
Eye of Observer
THE REFRACTING TELESCOPE

Cardboard tube
Eye lens
Sliding tube
Spectacle lens
A HOME-MADE TELESCOPE

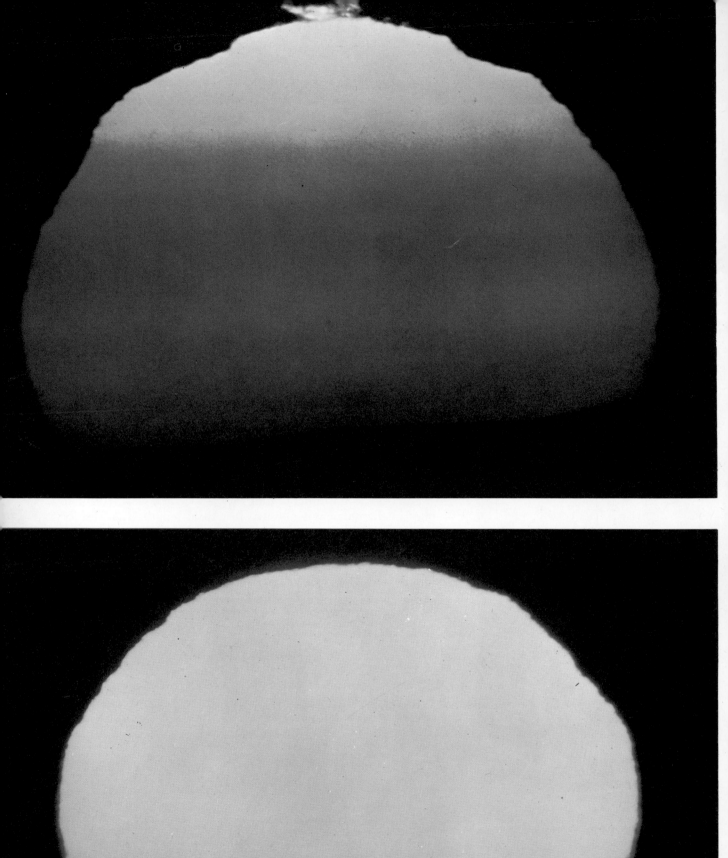

SOLAR DOMES AT PULKOVO. *The old Pulkovo Observatory at Leningrad, U.S.S.R., was destroyed during the war, and the present buildings are modern. The photograph, taken by Patrick Moore in 1960, shows domes housing equipment for studying the Sun*

10   the face of the Sun

PROJECTING THE SUN. *The only safe way to observe the Sun is to use a telescope to project the solar image on to a screen. This photograph was taken by W. M. Baxter in his observatory at Acton, London. The telescope is a 4-inch refractor*

◀ THE GREEN AND RED FLASHES. *These and other low-sun phenomena are due to the effects of the Earth's atmosphere. Photographs by Father D. J. K. O'Connell, S.J., Vatican Observatory*

GALILEO'S ASTRONOMICAL TELESCOPE allowed him to make a whole series of remarkable discoveries. It was in 1611 that he first saw that the Sun, which had always been regarded as a pure and unblemished globe, was not quite so perfect as had been believed. Here and there its dazzling disk was found to be disturbed by darker patches, now known as sunspots.

Such patches had been observed long before, by the Chinese. The ancient sky-watchers had no telescopes, but exceptionally large sunspots are visible without any optical aid when the Sun is shining through mist or light fog, so that its brightness is reduced. Galileo may not have been the first to see the spots telescopically. They were detected at about the same time by two other men— Christopher Scheiner, a German Professor of Mathematics, and Johann Fabricius, a young and promising observer who unfortunately died in 1616 when aged twenty-nine. However, the question of priority is not in the least important. The sunspots existed, and some way of explaining them had to be found.

Galileo himself was at first doubtful of the correctness of his observations, and did not publish them until 1612. His words were:

Having made repeated observations I am at last convinced that the spots are objects close to the surface of the solar globe, where they are continually being produced and then dissolved, some quickly and some slowly; also that they are carried round the Sun by its rotation, which is completed in a period of about one lunar month. This is an occurrence of the first importance in itself, and still greater in its implications.

The Church authorities particularly disliked the idea that the Sun might have spots on it, and Galileo became involved in a number of arguments.

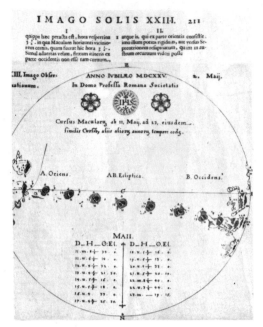

SUNSPOT DRAWING BY SCHEINER.
This series of observations was made in 1625 by Christopher Scheiner. Scheiner was one of the first telescopic observers to record sunspots, and there was an argument between him and Galileo with regard to priority—though in fact the question of priority is not in the least important

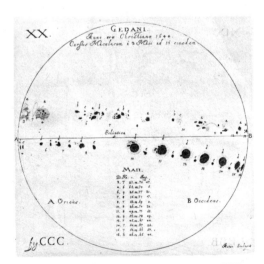

SUNSPOT DRAWINGS BY HEVELIUS.
Hevelius, one of the best telescopic observers of the seventeenth century, carried out most of his work from his private observatory at Danzig; his telescopes were among the best of their time, though very feeble judged by modern standards. His drawings of sunspots represented an advance on the work of Galileo, Scheiner and other pioneers

Modern observations have shown that the average distance of the Sun from us is approximately 93,000,000 miles. Since its apparent diameter can be measured, its real diameter can be worked out, and proves to be about 865,000 miles. This means that it is big enough to swallow up over one million bodies the size of the Earth; more accurately its volume is 1,300,000 times that of our world. The Sun's globe is much larger than the diameter of the Moon's orbit round the Earth.

Yet the Sun is not so massive as might be thought. If we could put it in one pan of a vast pair of scales, we would need only 332,958 Earths to balance it. In other words the Sun is not so dense as the Earth, and is only 1·4 times as dense as water—whereas on the same scale the density of the Earth is 5·5. Unlike the Earth, the Sun is not solid and rocky; it is made up of hot gas, and even at its surface it is fiercely hot. The temperature there is about 6,000° C, and near the centre of the globe it is believed that the temperature rises to well over 10 million degrees.

The fact that the Sun is so hot makes it a source of danger for the inexperienced amateur astronomer. To look straight at it through any telescope, or even a pair of field-glasses, is fatal. The concentration of heat upon the observer's eye will be enough to blind him permanently, and unfortunately this has happened to a large number of people. Even when the Sun is low down, and does not seem brilliant enough to dazzle the eye, the danger is still very real. Unfortunately it is possible to buy special darkened 'sun-caps', designed to fit over the eyepiece of a telescope for direct viewing of the Sun. These too are unsafe, and should not be placed on sale. A dark glass is always liable to splinter suddenly, and if this happens the observer may not be able to move his eye away in time.

There is only one sensible way to look at sunspots, and this is by the method of projection. First, the telescope is pointed at the Sun by sighting along the tube, making sure that a solid tin or cardboard cap covers the object-glass. The cap is removed, and the solar disk is projected on to a sheet of white paper or card. The Sun's face will be clearly seen, together with any spots which may be present.

There is one observation which may be made without any kind of optical aid. Under favorable conditions, a glorious spectacle may be seen at the moment of sunset—a flash of vivid green or red as the last portion of the Sun vanishes below the horizon. The effect is produced in the Earth's atmosphere, and is not a phenomenon originating in the Sun itself, but it is well worth looking for. It is best seen over a sea horizon.

Since the Sun's bright surface or *photosphere* is gaseous, and is not 'solid' in the ordinary meaning of the word, permanent features are not to be expected, and Galileo was right in saying that the sunspots are relatively short-lived. Some of them last for only a few hours, while the 'age record' seems to be held by a spot which persisted from June to December 1943, a period of nearly 200 days. However, this spot was not under continuous observation for the whole of that period. As Galileo realized, the Sun spins round on its axis, but instead of taking only twenty-four hours, as for the Earth, one rotation takes almost a month. What

Galileo did not know, and had no means of telling, was that the Sun does not spin as a solid body. At the solar equator the rotation period is 24·6 days, but at latitude 40° the period is 27·5 days, while at the Sun's poles the period has increased to as much as 34 days. This is an extra proof—if proof were needed—that the solar surface is made up of gas.

If you rotate a football which has specks of mud on it, the specks will seem to move round until they disappear over the edge of the football; if you go on spinning, the specks will reappear on the opposite edge. It is the same with sunspots. As the Sun rotates, the spots seem to be carried from one side of the disk to the other, finally vanishing over the edge or *limb*. If they persist, they will reappear over the opposite limb about a fortnight later. Their appearance will be affected, too; near the limb a spot is seen foreshortened, and will appear elliptical in form if the real shape is circular.

Suppose that we have a line of sunspots down the Sun's central meridian, as shown in the diagram. (Spots are never seen near the poles, but this makes no difference to the principle of the illustration.) 24·6 days later the equator will have completed one rotation, and the middle spot (A) will have come back to its original position, but in higher latitudes the rotation period is longer, and so the line of spots will no longer be straight. Evidently the Sun spins in a somewhat complex fashion. However, Galileo and Scheiner independently made estimates of the mean rotation period, and were not far from the truth.

Daily observations of the Sun, made by the projection method, will show the spot-shifts very clearly. Also to be seen are bright patches, generally associated with spots, which are known as *faculæ* (Latin: 'torches'). These faculæ may be regarded as luminous clouds lying above the solar photosphere; they often appear in positions where a spot-group is about to break out, and persist for some time after the group has disappeared. Consequently the appearance of faculæ on the Sun's eastern limb is often an indication that a spot is about to come into view from the far side.

As many early drawings show, a spot is not a simple dark blob. There is a central blackish part or *umbra*, and with larger spots a lighter surrounding *penumbra*. The shapes are often irregular, and one mass of penumbra may contain many umbræ. Yet even the darkest spot is not genuinely black. It is at a temperature of around 4,000 degrees (2,000 degrees cooler than the surrounding photosphere), and if it could be seen shining by itself it would

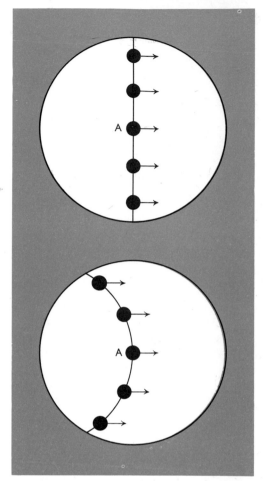

DIFFERENTIAL ROTATION OF THE SUN. *Region A, at the solar equator, completes its rotation quickest. In the polar regions, the length of the axial rotation period is appreciably longer*

SUNSPOT PHOTOGRAPHS. *The same sunspot group photographed on three different days; August 19, 1959 (left), 21 (centre) and 23 (right). Photographs by W. M. Baxter*

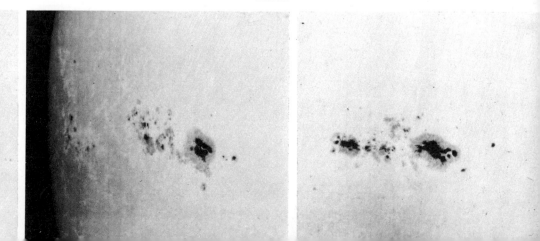

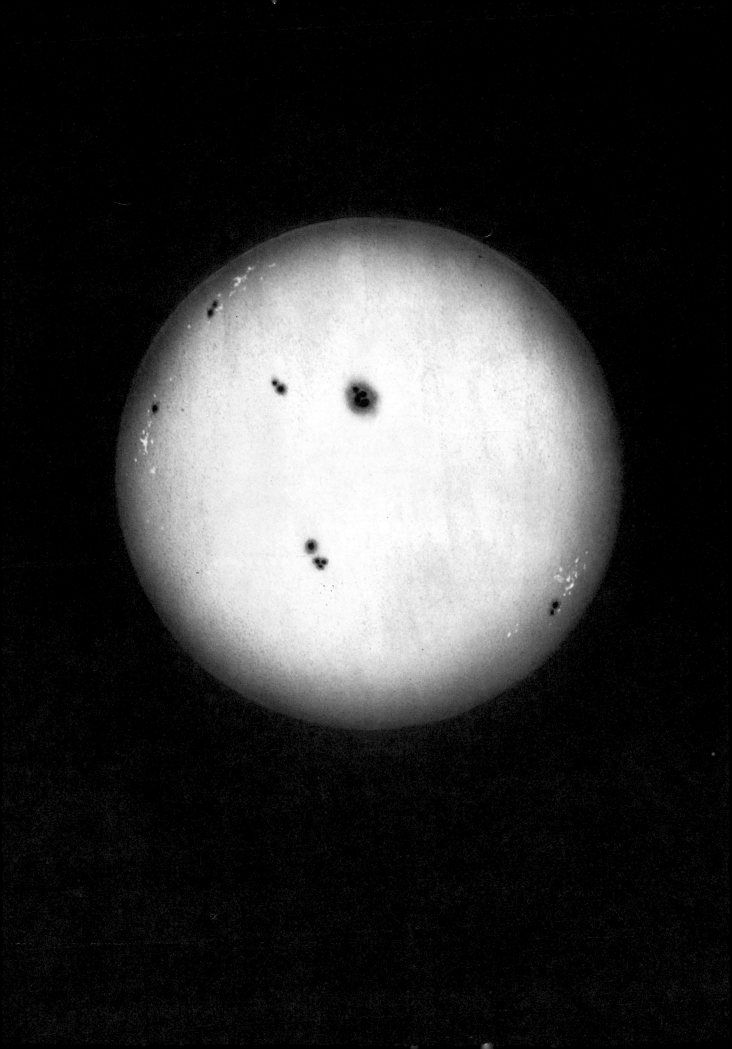

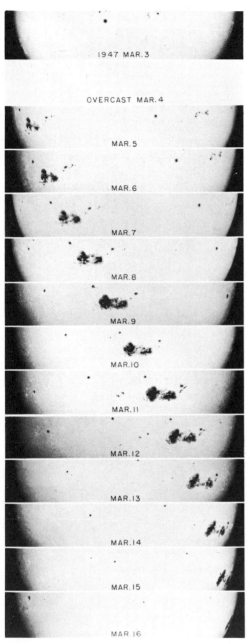

1947 MAR.3

OVERCAST MAR.4

MAR.5

MAR.6

MAR.7

MAR.8

MAR.9

MAR.10

MAR.11

MAR.12

MAR.13

MAR.14

MAR.15

MAR.16

THE GREAT SUNSPOT GROUP OF 1947. *A day-by-day record from March 3 to March 16. This was the largest spot-group observed since accurate records were begun. Photographs by Mount Wilson and Palomar Observatories*

THE SUN, *showing spot-groups and faculæ. The diminution in light near the edge of the disk 'limb darkening' is also apparent. From an observation made by Patrick Moore with a 4-inch refractor (projection method). Drawing by D. A. Hardy*

appear dazzlingly brilliant. It appears dark only by contrast with the still brighter background.

The Sun gives us as much light as 465,000 Full Moons would do. It is therefore difficult to justify the popular statement that a clear, moonlight night can be 'almost as bright as day'.

Galileo himself never followed up his work on sunspots. He was busy with other matters (including his troubles with the Church), and, as we have seen, his eyesight failed him towards the end of his life. We know now that when he first turned his telescope at the Sun, in 1610 and 1611, there were considerable numbers of spots; this was around the period of *solar maximum*. To understand what this means, it is necessary to jump forward two centuries to the work of Heinrich Schwabe.

Schwabe, born at Dessau in Germany in 1789, was an amateur observer. In 1826 he became interested in astronomy, and began to make daily drawings of sunspots. He carried on this work for many years, and in 1851 he made an important announcement. Every eleven years or so the Sun is particularly active, and there are many groups of spots; at such times there are generally at least four or five groups in view. Then the activity dies down to a minimum, and there may be periods when the Sun is completely clear. Subsequently, activity begins to build up once more, and after eleven years another maximum is reached.

This *solar cycle* is not perfectly regular, and the eleven-year period is only an average, but it is quite good enough for a general guide. Galileo was fortunate in that his early work was carried out near solar maximum. If he had started observing five or six years later he might never have found the spots at all, and by the time of the following maximum he had been forced to give up observing altogether.

Neither are all maxima equal in intensity. During the maximum of 1947–8, for example, some giant spots were seen; the greatest group ever recorded, that of April 1947, was easily seen with the naked eye. Very large spots, clearly visible without a telescope, were also seen at the maximum of 1958–9. The following maximum, that of 1969, was much less energetic than its predecessor. At present (1972) we are approaching minimum activity; the next maximum may be expected around 1979–80.

Sunspots are never visible near the Sun's poles, and in 1861 another German observer, Friedrich Spörer, discovered a curious 'law' which still bears his name. (Spörer was originally a schoolmaster, and took up astronomy as a hobby, though later in his life he worked at Potsdam Observatory.) During the early part of a solar cycle the spots appear some distance from the Sun's equator; but as the cycle progresses, the spots invade lower and lower latitudes. As the cycle draws to its end, and its groups die away, small spots of the new cycle start to appear in high latitudes once more. At solar minimum, therefore, two areas are subject to spots; near the equator, due to the last groups of the dying cycle, and in higher latitudes, due to the first spots of the new cycle.

Daily photographs of the Sun are now taken at various large observatories, and our records are very complete. Even before the end of Galileo's own century there was considerable information to go on. But astronomers have to admit that so far they

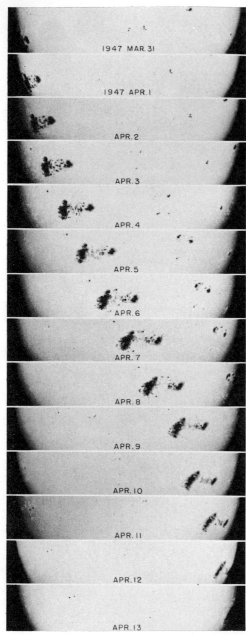

1947 MAR. 31

1947 APR. 1

APR. 2

APR. 3

APR. 4

APR. 5

APR. 6

APR. 7

APR. 8

APR. 9

APR. 10

APR. 11

APR. 12

APR. 13

THE GREAT SUNSPOT GROUP OF
1947. *A continuation of the record, from
March 31 to April 3—the rotation following
that shown on page 57. The two series of
photographs may be compared, showing the
changes in the spot-group. At its greatest
extent, in April, the group was easily visible
to the naked eye. It persisted for several more
rotations of the Sun before it finally disappeared*

do not know why sunspots appear; we know how they behave,
but we are very uncertain as to their real nature. Various theories
have been put forward, but our knowledge is still incomplete.
The Sun is an ordinary star, and it is very likely that other stars
show similar patches, but we cannot see them—simply because all
the other stars are so remote that no telescope yet built will show
them as anything but points of light.

Most of our modern knowledge of the Sun has been gained by
the use of more complex instruments based upon the principle of
the *spectroscope*, as will be seen later. All that Galileo, Scheiner and
their contemporaries could do was to record the various groups
and study their behavior. Meanwhile, attention was being paid
to an entirely different phenomenon which was only indirectly
connected with the Sun itself.

On the Copernican system—which, in spite of the Church, was
accepted by most scientists even before Galileo's death in 1642—
there are two planets, Mercury and Venus, which revolve round
the Sun at a distance which is less than that of the Earth. If therefore either of these planets passes directly between the Earth and
the Sun, it will be seen as a black spot passing across the solar disk,
and taking an hour or more to do so (the actual period depending,
of course, whether the planet passes right across the Sun, or close
to the edge of the solar disk). This is known as a *transit*. Obviously,
only Mercury and Venus can be seen in transit. The remaining
planets are farther from the Sun than we are—though it is worth
noting that anyone observing from, say, Mars would be able to
watch occasional transits of the Earth.

If Mercury and Venus moved round the Sun in the same plane
as the Earth, they would transit every time they reached inferior
conjunction. Unfortunately this is not the case. Mercury's orbit is
inclined to that of the Earth by 7 degrees, and that of Venus by
about 3½ degrees. These angles are small enough, but the result
is that at most inferior conjunctions both Mercury and Venus
pass either above or below the Sun in the sky, so that no transit
occurs.

It will be remembered that the last great work of Johannes
Kepler was the compilation of the Rudolphine Tables of planetary
movements. Kepler finished them in 1627, and from them he predicted that both Mercury and Venus would transit the Sun in 1631
—Mercury on November 7 and Venus on December 6. By that
time Kepler himself was dead, but the transit of Mercury was
successfully watched by the French mathematician Pierre Gassendi.

Mercury is a small world, and as it passed across the Sun in
transit it was too small to be seen without the aid of a telescope.
The transit of Venus promised to be more interesting. Afraid that
Kepler's prediction might be in error, Gassendi began watching
the Sun on December 4, and kept his observations continuously
until sunset on the 7th. To his great disappointment he saw
nothing. We now know the reason why. The transit occurred
indeed, but it took place during the northern night of December
6–7, when the Sun was below the horizon in France.

Kepler had forecast no more transits of Venus until 1761, but
fresh calculations were made by a young English clergyman, the
Rev. Jeremiah Horrocks, curate of Hoole in Cheshire, showing

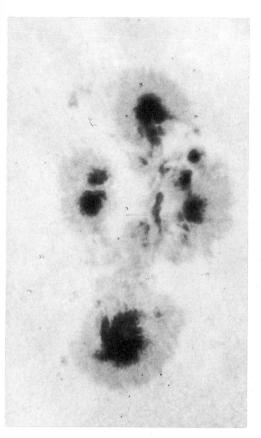

LARGE SUNSPOT GROUP, MARCH 31, 1960, *photographed with a 4-inch refractor by W. M. Baxter*

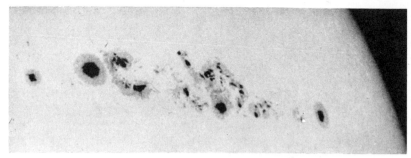

GIANT SUNSPOT STREAM, *February 20, 1956, photographed at Mount Wilson and Palomar Observatories*

that a transit would take place on November 24, 1639 (old style; the new style date is December 4—this occurred, of course, before the final change-over to our modern-type calendar). Horrocks finished his calculations only a short while before the transit was due, and he had time only to inform his brother Jonas, who lived near Liverpool, and his friend William Crabtree, who was close to Manchester. Both Jeremiah Horrocks and Crabtree were fortunate enough to see Venus in front of the Sun, though Crabtree's view was limited to a few minutes when the clouds luckily broke up.

Horrocks was an interesting man. He was born in 1619, studied at Cambridge, and became a firm supporter of the Copernican system. He seemed to be destined for a great scientific career, but he died in 1641 at the early age of twenty-two.

Transits of Mercury are not too infrequent; the last took place in 1970, and there will also be transits in 1973, 1986, and 1999. The more interesting transits of Venus are much rarer. They occur in pairs separated by an interval of eight years, after which over a century elapses before the next pair. Since Horrocks' time the only transits have been those of 1761, 1769, 1874 and 1882, while the next will be in 2004 and 2012. It is a pity that they are so infrequent, since at such times Venus appears large enough to be visible with the unaided eye. It cannot be confused with a sunspot, since it moves across the solar disk in a few hours, whereas a spot may be seen for a fortnight or so (provided that it lasts so long) before the Sun's rotation has carried it from one limb to the other.

It was tragic that Jeremiah Horrocks died so young. Had he lived, he might well have become one of the greatest astronomers in history. Even so, his name will not be forgotten; there is still an observatory near his old home which is named in his honor, and he will always be remembered as being the first man to see Venus as a black spot against the dazzling face of the Sun.

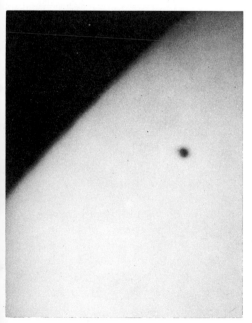

TRANSIT OF MERCURY, NOVEMBER 7, 1960, *photographed at 15 hours with the 10-inch reflector of the Norwich Astronomical Society*

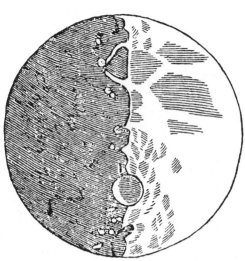

GALILEO'S MAP OF THE MOON. *This was the second lunar chart drawn with the aid of the telescope; the first was Harriot's. It is naturally rough, but various features are identifiable*

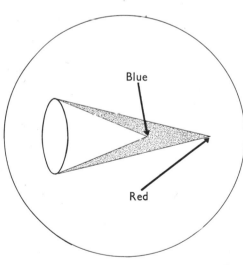

UNEQUAL REFRACTION OF LIGHT. *Blue light is bent or refracted more than red, and is brought to a different focus. In the diagram, the differences have been considerably exaggerated, for the sake of clarity*

GALILEO DIED IN 1642. Isaac Newton whose work may be said to have laid the foundations of present-day theoretical astronomy, published his greatest work—the *Principia*—in 1687. Before considering the life of Newton and the laws of gravitation, something should be said about the other astronomers of the period, who were busy using telescopes to explore the Earth's nearest neighbors, the bodies of the Solar System.

The first Britons to look at the Moon through a telescope seem to have been Thomas Harriot, who drew up a lunar chart, and Sir William Lower, who lived at Traventy in Pembrokeshire. Lower deserves to be remembered for his graphic description of the Moon's surface. He said that when seen through the telescope it resembled a tart which his cook had made—'here some bright stuff, there some dark, and so confusedly all over'! In fact, this is not really a bad description of the Moon as seen through a low-powered instrument which will not show anything sharply. Lower seems to have made no really serious observations.

Observers began to draw maps of the Moon: Scheiner produced one in 1614, Gassendi another in 1640, and so on. It was in 1647 that the first reasonably useful chart was produced by Johann Hevelius, a rich merchant who was also a City Councillor of the Baltic seaport of Danzig.

Hevelius was born in 1611, and went to Leyden University. After he had completed his education he spent three years travelling around England, France and Italy, after which he returned to Danzig. He gave all his spare time to astronomy, and he equipped what was then the best observatory in Europe. It was placed on the roof of his home—which, from all accounts, consisted of four separate buildings—and of course it contained telescopes. Yet these telescopes were quite unlike modern instruments. They had small object-glasses, but they were immensely long.

There was a good reason for this. As all the early observers found, a refracting telescope of the kind built by Galileo had the unfortunate effect of producing 'false color' round any bright object. A star, for instance, would seem to be surrounded by colored rings. This was both inconvenient and a handicap to proper observing, and at the time nobody knew what was the cause.

As was found later, the root of the trouble was in the fact that what is normally called 'white light' is not white at all; it is a blend of all the colors of the rainbow, from red to violet. When a beam of light enters an object-glass, and is refracted, the various colors are not bent equally. Red light, for instance, is refracted less than blue, and so is brought to focus at a different point. This is shown in the figure. The diagram is out of scale, since the real difference between the red focus and the blue focus is extremely small—a tiny fraction of an inch. However, the effect was enough to make the first refractors very inefficient.

150-FOOT TELESCOPE USED BY HEVELIUS. *One of the clumsy 'aerial telescopes' of the seventeenth century*

Even Newton was puzzled, and it was not until many years later that a partial remedy was discovered by an English amateur astronomer named Chester More Hall. Meanwhile Hevelius and others found that the false color trouble could be reduced by making telescopes of very long focal length.

This seemed to be the only solution, and Hevelius' best telescope had a focal length of 150 feet! It was hopeless to make a tube of such dimensions, and so the object-glass had to be fixed to a mast 90 feet high. When we look at old drawings of the instrument, we wonder how Hevelius managed to use it at all; it must have been remarkably clumsy and awkward, but he made good observations with it, and even produced a star catalogue. (This instrument was

QUADRANT MADE AND USED BY HEVELIUS, *dated 1659. From an old woodcut*

LUNAR MAP BY RICCIOLI. *Riccioli's lunar map appeared in 1651; it was based mainly upon observations made by his pupil Grimaldi. His system of nomenclature soon superseded the rather poor geographical names of Hevelius*

not the longest of the 'aerial telescopes'. Christiaan Huygens used a 210-foot refractor, and the French astronomer Adrien Auzout is said to have designed a 600-foot telescope, though it was never built.)

Hevelius is best remembered because of his catalogue of comets, which was the best of its time, and because of his work in connection with the Moon. He seems to have realized that the dark areas are plains and that the brighter areas are mountainous, and he made measures of the peak-heights which were better than Galileo's. He also named the various features on the Moon's surface, and drew a complete map just under one foot in diameter.

Hevelius' scheme was to give Earth names to the lunar craters and plains. One large crater was named by him 'Etna'; another crater was 'the Greater Black Lake', and so on. The system was not convenient, and was soon abandoned. Less than half a dozen of Hevelius' names are still used.

Meanwhile a Jesuit Professor of Mathematics, Francisco Grimaldi, had been observing the Moon in order to draw up a map. The map was actually published by another Jesuit, Giovanni Riccioli, who was considerably older than Grimaldi (he was born in 1598) and taught first at Padua and then at Bologna. As a theorist Riccioli was by no means outstanding, and he rejected the Copernican system, preferring to believe Tycho's idea that the Sun went round the Earth. However, he was a reasonably correct observer, and although his map was based on Grimaldi's studies it is possible that he made some contributions himself.

After careful thought he decided to reject Hevelius' names, and

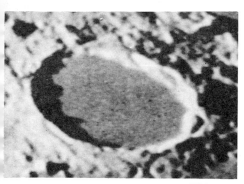

THE LUNAR CRATER PLATO. *This was the formation which Hevelius called 'the Greater Black Lake'; its floor is very dark in hue, and is relatively smooth. The full diameter is 60 miles. The crater is circular, but appears elliptical due to foreshortening effects. Any small telescope will show it, and the darkness of its interior means that it is recognizable under all conditions of illumination*

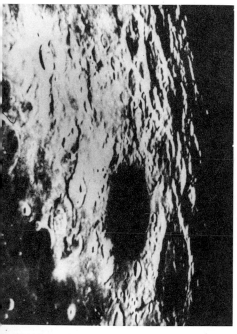

LUNAR CRATERS: REGION OF GRIMALDI AND RICCIOLI. *Grimaldi is the very large formation with an almost black floor; indeed, this is the darkest point on the Moon's surface. The walls of the crater are relatively low, but Grimaldi is always easy to recognize; had it been placed nearer the centre of the Moon's visible disk it might well have been classed as a minor 'sea'. Riccioli, below and to the right of Grimaldi, is rather smaller, but has a diameter of about 100 miles; its floor contains one very dark area. Note the interesting 'valley' entering Riccioli from the west (left), to which attention has been drawn by the Japanese observer Miyamori*

to use a completely different system. The dark plains, which he thought to be seas, were given attractive names such as the Ocean of Storms, the Sea of Showers and the Sea of Serenity. Naturally Riccioli used the Latin versions, so that the Ocean of Storms became 'Oceanus Procellarum', the Sea of Showers 'Mare Imbrium', and so on. (Astronomically, we still retain the Latin forms.) The craters were named in honor of famous men and women, usually those who had been connected with science in some way.

Since Riccioli was an admirer of Tycho Brahe, and believed in Tycho's system rather than that of Copernicus, he gave the great Danish observer the most prominent crater on the Moon—a 56-mile formation in the south, which is the centre of a system of extraordinary bright streaks or *rays*. Copernicus, together with Aristarchus—who had advanced the 'moving Earth' theory so long before—was, as Riccioli tells us, 'flung into the Ocean of Storms', but at least both the selected craters are very conspicuous. Copernicus, like Tycho, is the centre of a ray system, while Aristarchus is the brightest object on the lunar surface.

All the large craters were renamed. Hevelius' 'Greater Black Lake' was christened in honor of the Greek philosopher Plato; Julius Cæsar and Sosigenes were given less prominent craters near the centre of the Moon's disk; Galileo is to be found towards the edge of the Ocean of Storms, and so on. Needless to say Riccioli took care to name a large crater after himself, and Grimaldi was allotted an even bigger formation close by.

Riccioli's system is still in use. Later astronomers added to the list, and the latest lunar maps, based upon measures made by photography, include over 800 separate names. Some of them are rather unexpected. For instance, there is a Birmingham on the Moon, named after a nineteenth-century Irish amateur, John Birmingham. We also meet with Billy (after Jacques Billy, a French Professor of Mathematics who lived from 1602 to 1679) and even Hell (after Maximilian Hell, of Hungary, who is best remembered for his work on the 1769 transit of Venus).

In some ways the system has not been well applied. It is true that Ptolemy, Eratosthenes and other great men of the past have been given suitable craters, but the partly ruined object which has been named in honor of Newton is by no means easy to identify, though it is in fact the deepest crater on the Moon, and has walls rising to more than 30,000 feet above its floor. Neither do the craters called after Galileo, Halley (of comet fame), Horrocks and others do justice to the names they bear. However, the general scheme will certainly never be altered now. It has been in use for over three centuries, and has become firmly established. (Now that the far side of the Moon has also been mapped, names have been allotted to the features there also, and the general plan has been followed.)

No better maps of the Moon were drawn for well over a hundred years after Riccioli's time. Hevelius was unfortunate; his elaborate roof-top observatory was burned down, together with many of his unpublished observations, and it is said that the copper-plate of his famous Moon map was afterwards made into a teapot. But the greatest observer of the period was undoubtedly a Dutchman,

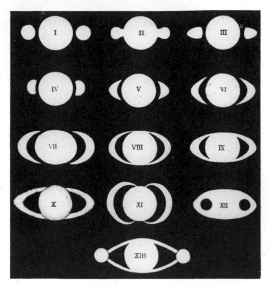

DRAWINGS OF SATURN MADE BY HUYGENS

THE CHANGING ASPECTS OF SATURN'S RINGS ▶

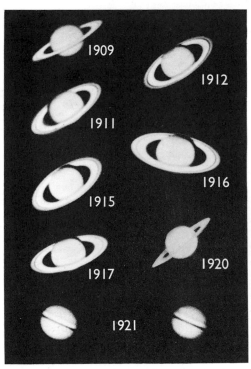

PHOTOGRAPHS OF SATURN, *showing the ring-system under different angles. Lowell Observatory photographs*

Christiaan Huygens, who turned his attention to the planets.

Huygens was a remarkable man. He was born in 1629, and studied at Leyden. For some years he lived in Paris, and came to England, where he met Isaac Newton. He became an expert telescope-maker, and invented a new type of eyepiece which we still term the 'Huygenian'. More important still were the improvements which he made in the construction of watches, and his invention of the pendulum clock. Galileo, as we have seen, had paid some attention to the pendulum as a timekeeper, but to Huygens goes the honor of being the first to build a clock on this principle.

Galileo had been badly puzzled by the curious appearance of the planet Saturn. He had suggested that the globe was triple, but when the two attendant bodies disappeared he was completely at a loss. Huygens' telescopes were more powerful, and in 1655 he solved the problem, though he did not announce his discovery until four years later.

Huygens saw that Saturn is not triple at all. The globe is surrounded by what he calls 'a thin, flat ring, nowhere attached to the body of the planet'. The ring is circular, though seen from the Earth it naturally appears elliptical; it measures 175,000 miles from side to side, but it is probably no more than ten miles thick.

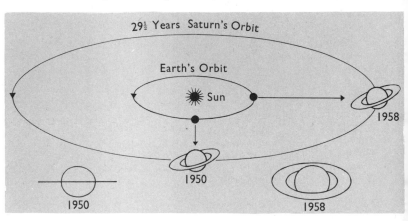

This means that when the ring is edge-on to us, as happens every fifteen years or so, it can be seen only as a thin line of light. Indeed, it disappears except in telescopes of some size.

The drawings show the changing appearance. In 1958, when the ring-system was at its greatest angle to us, the planet was a glorious sight; in 1966 the rings were again edge-on, but are now easily visible once more. No wonder that Galileo was mystified.

It was natural to think of the ring as being a solid sheet of material, but we now know that its real nature is quite different. Saturn is a massive world and has a powerful gravitational pull, so that a solid or liquid ring would quickly be torn to pieces. It is composed of a large number of relatively small particles moving round the planet in the manner of dwarf moons. These particles are too small to be seen separately from a distance of over 700 million miles, which explains the 'solid' appearance. Some years after Huygens' announcement, his colleague Cassini discovered a gap in the rings which is still known as 'Cassini's

SATURN. *From an observation by Patrick Moore. Drawing by D. A. Hardy*

HUYGENS' ORIGINAL DRAWING OF MARS. *This, the first telescopic drawing of Mars to show surface detail, was made by C. Huygens in 1659. The Syrtis Major, the most prominent of the dark features, is clearly recognizable*

JUPITER. *A photograph taken in red light with the 200-inch Hale reflector at Palomar. The Red Spot is shown. The black disk above is the shadow of Jupiter's third satellite, Ganymede, and Ganymede itself is seen to the right of the planet*

Division', and in 1848 a third much dimmer ring was found, closer to the planet than the bright pair.

So far as we know, Saturn is unique in the heavens. It may be that the rings represent the wreck of an old satellite which happened to wander so close to the planet that it was torn to pieces by Saturn's gravitational pull; it may be that there is another explanation. At any rate, the ring-system forms what may be regarded as the most beautiful spectacle in the sky.

At about the same time Huygens made another interesting discovery. Jupiter was known to have four moons, and now it was found that Saturn had one of its own. Titan, as the new satellite was named, proved to be decidedly large, since it has a diameter of well over 3,000 miles. It is in fact bigger than the planet Mercury, though not so massive. Moreover it has an atmosphere, though this was not discovered until as recently as 1944.

Though Titan is actually larger than any of Jupiter's moons, it is almost twice as remote from us, and appears much fainter. Any modern 3-inch refractor will show it, but Huygens must have needed keen eyes to detect it with the clumsy 'aerial telescope' which he had to use.

Saturn is quite unlike the Earth. Instead of being solid and rocky it is composed of cold gas, mainly hydrogen and hydrogen compounds. Huygens could not know this, but he soon found that Mars at least was much more Earthlike. In 1659 he drew it from telescopic observations to show a V-shaped marking known today as the Syrtis Major. Probably it is due to living organisms of some kind, and by watching its apparent drift across the planet from one limb to the other Huygens was able to tell that Mars rotates on its axis in about twenty-four and a half hours. The 'day' there is about half an hour longer than on Earth, so that Huygens' original estimate was rather too short.

Huygens made these discoveries when he was living in Holland, but in 1665 he went to France, at the invitation of Louis XIV, and stayed there for sixteen years. It was during this period that he completed his greatest work, the invention of the pendulum clock. Unfortunately there were religious troubles to be faced; Huygens was a Protestant, and in 1681 all Protestants in France became unpopular. Huygens accordingly returned to his native country, and died at The Hague in 1695.

Another astronomer busy in France at this time was Giovanni Domenico Cassini. Cassini was born in Italy in 1625, and in 1650 became a Professor at Bologna. Like Huygens he was a careful observer of the planets; he measured the rotation period of Mars, and also that of Jupiter, which was known to be flattened at the poles. This fact is particularly interesting. As Cassini found, the diameter of Jupiter is 88,700 miles if measured through the equator, but only 82,500 if measured through the poles. The cause lies in the planet's quick rotation. Instead of being about twenty-four hours long, the 'day' there is less than ten hours long, but as with the Sun the period is shorter at the equator than at the poles.

In 1666 Cassini saw that the two poles of the planet Mars are covered with whitish deposits which look very like snow or ice. Until very recently it was thought that they must be thin caps

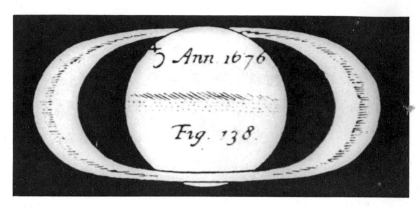

CASSINI'S DRAWING OF SATURN. *This is one of the drawings made in 1676 by G. D. Cassini, at Paris. It shows the famous division in the rings which Cassini discovered, and which is still known by his name. The drawing is naturally rough by modern standards, and the shape of the ring-form is not accurate, but it was of course made with one of the clumsy 'aerial telescopes', and it is a tribute to Cassini's skill as an observer that the rings are shown in recognizable form. One of the belts crossing the surface of the planet is also shown. Belts on Saturn resemble those of Jupiter, but are considerably less prominent*

similar in composition to those of the Earth, though the recent evidence drawn from space-probes indicates that they are more likely to be made up of solid carbon dioxide.

It was in 1666, too, that the French astronomer Adrien Auzout pointed out that it was time to establish a national observatory equipped with the best instruments available. Up to this time there had been only one such establishment—at Copenhagen, which had been completed in 1656 and was burned down during the following century. All the great observers, such as Galileo, Hevelius and Huygens, had built and set up their own equipment. Auzout wanted to alter all this, and fortunately the French King, Louis XIV, was a patron of science. The result was the founding of the famous Academy of Sciences in Paris, and no time was lost in starting the construction of an observatory. Cassini seemed to be the obvious man to direct it, and he accepted an invitation to come to France.

Difficulties arose almost at once, mainly because the King wanted his observatory to look magnificent and Cassini was more interested in its scientific value. As soon as Cassini arrived and saw the half-finished building, he told Louis that unless it was drastically altered it would be of no use whatever. Louis was far from pleased, and a serious quarrel was only narrowly avoided. Eventually Cassini erected his instruments in the open air outside the observatory. This was by far the best plan, since he was still forced to use 'aerial telescopes' of small aperture and enormous focal length. However, it was at least a beginning, and ever since then the Paris Observatory has been one of the most important in the world.

The French skies are less clear than those of Italy, but Cassini was able to carry on his work. He discovered four more satellites of Saturn, now known as Iapetus, Rhea, Dione and Tethys; in 1675 he found the famous Division in Saturn's ring-system; he fixed the rotation period of Mars at 24 hours 40 minutes, which is only three minutes too long, and he was almost as successful in the case of Jupiter. Even more remarkable was his work in connection with the distance of the Sun. Kepler had believed the Sun to be only 14,000,000 miles away from us, but Cassini amended this to 86,000,000 miles. Since the real value is 93,000,000 miles, he was not very wide of the mark. He never went back to his homeland, and when he died in 1712 his son Jacques Cassini succeeded him as director of the Paris Observatory.

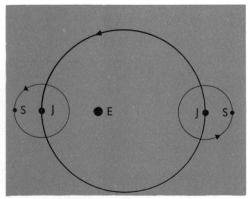

MEASURING THE VELOCITY OF LIGHT. *This shows the method adopted by Ole Rømer. When Jupiter (J) is at its closest to the Earth (E), the light from the satellite (S) has a lesser distance to travel, and so its eclipses are earlier than had been predicted*

APPARENT MOVEMENT OF MARS AMONG THE STARS, 1960. *At this time Mars was near opposition, and the constellation of Gemini, with the 'twins' Castor and Pollux, is shown; the shift of Mars is very obvious. The dates of the photographs are:* (upper) *October 26,* (right) *November 19,* (lower) *December 26. Color photographs taken by K. S. G. Stocker*

One of Cassini's colleagues at Paris was a Dane, Ole Rømer, who was responsible for yet another great advance—the measurement of the speed of light.

Cassini had made careful measures of the satellites of Jupiter, and had produced tables of their movements which were much the best in existence. In particular it was possible to work out when the four moons should be eclipsed by Jupiter's shadow, just as our Moon is sometimes eclipsed by the shadow of the Earth. However, he found that his predictions were often wrong; at times the eclipses occurred too early, at other times too late. There seemed to be nothing wrong with the tables of movements, and the eclipses of the satellites were not difficult to observe. Anyone who is equipped with a small telescope can watch them and can time them accurately.

Cassini wondered whether the speed of light might have something to do with the errors, but he did not follow up the idea. Rømer did. He discovered that eclipses occurred too early when the Earth was at its closest to Jupiter; when Jupiter and the Earth were farthest apart, the eclipses of the satellites were too late—simply because the light from them had taken longer to reach us, as shown in the diagram. Rømer calculated that instead of moving instantaneously, light travels at a rate of 186,000 miles per second. This was an excellent result; and is very close to the value fixed by modern methods.

Rømer's work did not end here. In 1681 he went back to Denmark and became Professor at Copenhagen, where he was responsible for various improvements in astronomical instruments, leaving Cassini to continue observing at Paris.

Clearly, then, observational astronomy had made great strides since Galileo had first turned his tiny telescope to the heavens in the winter of 1609. But theoretical work had been making even more striking progress, and this brings us to the greatest figure in the whole history of astronomy—Isaac Newton.

12   the genius of Newton

COPERNICUS HAD TAKEN the first real step in working out a correct picture of the universe. Kepler, Galileo and others had proved that the Copernican theory was correct, and that the Sun, not the Earth, lies in the centre of the planetary system. Yet many problems remained, and it was left to Isaac Newton to put astronomical science upon a really firm footing.

Newton was born at Woolsthorpe, near Grantham in Lincolnshire, in 1642. His father died before he was born, and his mother was left in charge of the family farm. There was little money to spare, and moreover England was in a disturbed state, since this was the time of the civil war between King Charles I and Cromwell's Roundheads. Isaac went to the village school, but showed no signs of unusual intelligence. All we know of his early school career is that he was particularly fond of making models.

When he was twelve he was sent to King's School, Grantham, which had been founded over a century earlier by Henry VIII. For his first few terms he remained near the bottom of his class. Then—according to a story which may or may not be true—a bigger boy, who was above Newton in form, jeered at him and kicked him. Newton had no intention of bearing such treatment, and a fight followed, ending only when Newton had beaten his rival and rubbed his nose against the wall. From that time on he worked hard, and reached the top position in the school.

Meanwhile his mother had married again, but in 1656 her second husband died, and Mrs. Newton brought Isaac back from Grantham to help her run the farm. The experiment was not successful. The boy was not in the least interested in farming; it is said that when he was supposed to be keeping an eye on the laborers he used to spend his time sitting behind a hedge, working out problems in mathematics. Wisely his mother sent him back to Grantham, and in 1661 he went to Cambridge University, enrolling as an undergraduate at Trinity College. Here he met Professor Isaac Barrow, who became his tutor in mathematics.

Barrow was the son of a linen draper. Like Newton he had done little work during his first years at school, and had distinguished himself only by his fondness for fighting. As he was very strong, and won far more fights than he lost, he was held in respect by the other boys. Later he went to Cambridge, and became a Fellow of Trinity College. During the Civil War he left England, and

STATUE OF NEWTON. *This photograph, taken by Patrick Moore in 1961, shows a statue of Newton which has been set up in the Lincolnshire town of Grantham. It was in this town that Newton received his education, and his birthplace, Woolsthorpe, is not far off. It was in Lincolnshire, too, that Newton carried out much of his important work—notably during the Plague period, when Cambridge University was closed, and all the students had been sent back to their homes*

PRODUCTION OF A SPECTRUM. *When a beam of apparently 'white' light is passed through a prism, it is split up into its component colors, and in this color photograph the production of a rainbow or continuous spectrum is shown. This effect was first noted by Newton*

A 12·5-INCH NEWTONIAN RE-FLECTOR. *This photograph shows a typical Newtonian reflector on an equatorial mounting. The focal length is 110 inches. The telescope was used by an English nineteenth-century observer, W. F. Denning, for his studies of planetary surfaces. It is now installed at Henry Brinton's private observatory at Selsey, in Sussex. Photograph by Patrick Moore*

THE 120-INCH LICK REFLECTOR. *This is one of the most modern of large reflectors, and is one of the largest telescopes in the world; it is surpassed only by the Palomar 200-inch. The tube is a skeleton, and the large size of the instrument means that various optical systems can be used*

travelled around Europe and Asia Minor. Although he was now a well-known scholar, he still kept his love of fighting, and on one occasion he showed his courage very plainly. During a sea voyage from Leghorn to Smyrna his ship was attacked by pirates. Barrow was not to be frightened; he stayed on deck, and fought so bravely that the pirate vessel sheered off.

Barrow was a Royalist, and came back to England after Charles II returned to the throne in 1660. In 1663 he was appointed to the 'Lucasian Chair' of mathematics, so called because the money to found the appointment had been provided by the will of a Mr. Lucas.

Barrow was a strong-minded man as well as a clever one, and he was the ideal tutor for the rather shy and retiring Newton. The two worked very happily together. Newton took his scientific degree in 1665, and it is worth noting that in 1669 Barrow resigned the Lucasian Chair of mathematics so that Newton could succeed him. Clearly, then, he had great faith in his pupil's ability, and this faith was more than justified in later years.

Fresh disasters had come to England. In 1665 the Great Plague struck London; 17,000 people died during August of that year, and 30,000 during September. The King and his Court left London, but cases of plague were reported from other parts of the country, and there was no real safety anywhere. When the disease reached Cambridge the authorities wisely closed the University. All the students were sent back to their homes, and Newton accordingly returned to Woolsthorpe.

To most men such an interruption would probably have been a handicap, but to Newton it was an advantage. He was able to work quietly, on his own, and with no money troubles. For the next year or so he was left in peace, and he managed to do a tremendous amount of research.

One subject which interested him was the nature of light. We have seen that the object-glasses of early refractors produced false color, and that this is due to the unequal bending of the different colors which blend together to make 'white' light. Newton demonstrated this by an ingenious experiment. He passed a beam of sunlight through a prism, and split it into the usual rainbow. He then passed one particular color—violet, for instance—through a second prism. No second spectrum was produced, which proved that the violet light was not a blend of different colors.

Newton's aim was to produce an *achromatic* object-glass—that is to say, a lens which would not produce false color. Unfortunately he could see no way of doing it, and after considering the matter carefully he decided that there was only one solution. He must build a telescope which did not need an object-glass at all.

Some years earlier, in 1663, a Scottish mathematician named James Gregory had suggested using a mirror to collect light instead of a lens. Gregory never built such an instrument—as he himself admitted, he had no practical skill—but Newton developed the idea, and produced the first *reflector*. His arrangement was not exactly the same as Gregory's, but it was more convenient, and is much more widely used today.

With Newton's arrangement, the light from the Moon (or whatever object is to be studied) passes down an open tube until it hits a mirror at the lower end. This mirror is curved, and reflects the light back up the tube, directing it on to a smaller mirror or *flat* placed at an angle of 45°. The flat sends the rays into the side of the tube, where they are brought to focus; the image is magnified by an eyepiece in the normal way. In a Newtonian reflector, then, one looks into the tube instead of up it.

Since a mirror reflects all parts of the spectrum equally, the false color trouble does not arise, and Newton's first instrument was very successful. It had a metal mirror 1 inch in diameter, and was presented to the Royal Society of London in 1672. It caused a major sensation, and led to quarrels between Newton and some of his fellow-scientists who disagreed strongly with his views about the nature of light.

Newton's reflector looks very small when compared with modern instruments. Today many amateur observers have reflectors with mirrors 6, 8, 12 or even 18 inches in diameter—made not of metal, but of glass coated so as to give high reflectivity. Moreover many amateurs make their own mirrors, grinding them carefully into the correct optical shape. Neither is it necessary to have a

NEWTON'S FIRST REFLECTOR. *The photograph shows a replica of the first reflecting telescope, demonstrated by Newton to the Royal Society. The mirror, 1 inch in diameter, was made of speculum metal. Though it was so small, the telescope functioned perfectly, and demonstrated the soundness of Newton's arrangement*

THE MOUNT WILSON 60-INCH REFLECTOR, *seen from the north-west. This was the first of the large Mount Wilson reflectors, made by Ritchey at the instigation of George Ellery Hale; it is interesting to make a comparison between this vast instrument and the original small telescope made by Newton. Large reflectors such as the 60-inch are seldom used for visual observations, and nearly all their work is carried out by means of photography*

AN 8·5-INCH NEWTONIAN REFLEC-
TOR. *This has an equatorial mounting made
in the late nineteenth century by Browning.
The focal length is 52 inches. Photograph by
Patrick Moore*

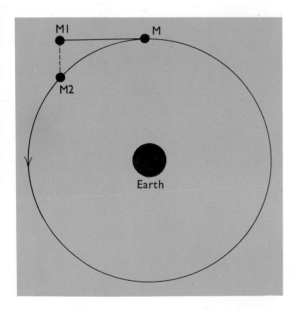

GRAVITATIONAL EFFECTS OF THE
EARTH UPON THE MOON. *This dia-
gram, which is not to scale, shows the Moon's
path round the Earth. But for the presence of
the Earth, we may suppose that the Moon
would move from M to M1 in one minute.
However, the Earth's gravitational pull
means that instead of moving uniformly from
M to M1, the Moon is 'pulled down' to M2.
It may be said that the Moon has 'fallen'
from M1 to M2 in one minute, and it goes on
'falling' all the time, though it does not drop
any closer to the ground*

solid tube; with many reflectors the 'tube' takes the form of a skeleton framework rigid enough to hold the large mirror and the flat firmly in position. In professional observatories, really giant reflectors have been set up. The largest in the world at present, the Palomar telescope in America, has a mirror 200 inches across, while the Russians are now testing an even bigger instrument with a 236-inch mirror.

On Gregory's plan the light was to be reflected back down the tube by a smaller curved mirror, as shown in the diagram on page 73, and passed to the eyepiece by way of a hole in the main mirror. This is also the case with the 'Cassegrain' arrangement. Many such reflectors exist today.

Yet Newton's principle is so convenient, and so sound, that it is by far the most popular, at least for telescopes of less than 18 inches aperture. In some ways there can be no doubt that a reflector is better than a refractor; not only is there no false color, but a mirror is much cheaper and easier to make than a lens of equal light-gathering power. Naturally there are disadvantages also, but the ideas put forward by Newton still hold good in modern astronomy.

Newton's name will always be associated with the laws of gravitation, and this research too was begun at the time when Cambridge University was closed because of the Plague danger. This brings us to the story of the falling apple—which is particularly interesting because it seems to be true.

According to the tale, Newton was sitting in his Woolsthorpe garden one afternoon when he saw an apple fall from a tree branch to the ground. He began to wonder why the apple had fallen. There must be some definite force which pulled it to the ground; but what was this force, and how far did it extend? Gradually Newton began to see that the force which pulled on the apple was the same as the force which keeps the Moon in its path round the Earth—or the Earth in its path round the Sun. This led him on to the idea of *universal gravitation*, according to which every particle of matter attracts every other particle with a force which becomes weaker with increasing distance.

Put in this way, the story is over-simplified; it must have taken Newton a long time to come to his decision and many months of hard work to express it in proper mathematical form. But the apple was the starting-point, and it will be helpful to follow the reasoning further.

Suppose then instead of being 20 feet or so in height, the tree had been 200 feet, or even 200 miles? The apple would still have dropped to the ground, gathering speed as it fell. This would also apply to a tree 239,000 miles in height. The Moon is 239,000 miles away from us—so why does it not fall, just as the apple did?

Newton found the answer. The reason why the Moon does not drop is because it is moving. It is not easy to give a correct every-day analogy, but some idea of what is meant can be gathered from taking a cotton-reel and whirling it round on the end of a string. The reel will not fall so long as it continues moving quickly enough for the string to remain tight, and the Earth's pull on the Moon may be said to act in much the same way as the string on our cotton-reel.

Now suppose that the man holding the string lets go suddenly. The reel will then fly off in a straight line. If we neglect the pull of the distant Sun, we can see that the Moon too would move off in a straight line if the Earth were not pulling upon it, and Newton realized that any moving body will continue its motion in a straight line unless some outside force is acting on it. This is the famous *law of inertia*.

Newton knew the force of the Earth's pull at ground-level, since this was the force affecting the apple, and he found the law according to which the force should weaken with increasing distance from the Earth. According to his calculations the Moon should 'fall' 15 feet per minute, which would of course be the distance between M1 and M2 in the diagram.

This did not agree with observation; the distance 'fallen' in one minute is not 15 feet, but only 13 feet. As Newton said, the figures 'agreed pretty nearly', but not well enough to satisfy him.

In this sort of calculation, a body such as the Earth behaves as though all its mass were concentrated at a single point at the centre of the globe. As Newton had to make his observations from the Earth's surface, he had to know the distance from the surface to the centre of the globe. In other words, he had to know the value of the Earth's radius. There is a story that the 2-foot error in the fall of the Moon was due to Newton's having used a wrong value for the Earth's radius. This story, however, is not true; one link in the mathematical argument was still missing, and it was not until years later that Newton found out what was wrong. Meanwhile he found that in order to make his calculations he had to develop an entirely new branch of mathematics. He called it the 'method of fluxions', but it corresponds to what we now term the *calculus*.

Newton returned to Cambridge after the end of the Plague danger, and in 1672 he was elected to the Royal Society, a scientific body which had been founded early in the reign of Charles II. He was still comparatively unknown; he had said nothing at all about his researches into gravitation, and his first official contribution to astronomy was his reflecting telescope. Unfortunately he had an immediate disagreement with another Fellow of the Royal Society, Robert Hooke, who was to play an important part in Newton's life.

Hooke was seven years older than Newton, and a man of different type. He was physically weak, and it is said of him that 'his figure was crooked, his limbs shrunken; his hair hung in dishevelled locks over his haggard countenance. His temper was irritable, his habits solitary.' He was violently jealous and suspicious, and frequently claimed the credit for work which had been done by others. This is not to say that Hooke was dishonest. He was in fact perfectly well-meaning, and there can be no doubt that he was brilliantly clever. His activities spread into all branches of science. He invented the 'universal joint' known to every mechanic, and also made some meteorological instruments, such as the hygrometer for measuring the wetness of the air. On the other hand he studied so many things that he seldom followed any particular investigation through to the end, as Newton did, and though he was an excellent mathematician he was by no means Newton's equal.

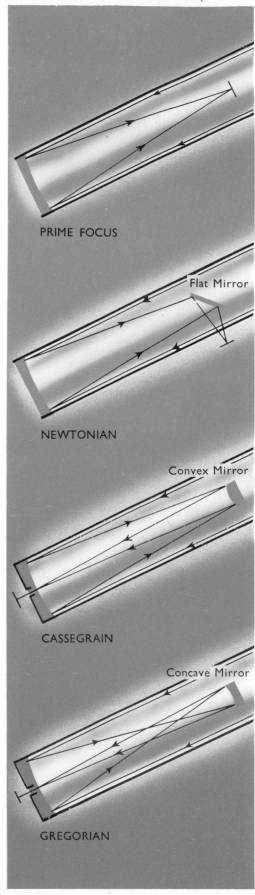

PRIME FOCUS

Flat Mirror

NEWTONIAN

Convex Mirror

CASSEGRAIN

Concave Mirror

GREGORIAN

FOUR TYPES OF REFLECTORS

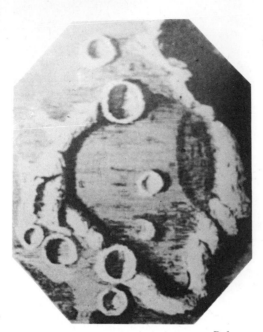

LUNAR DRAWING BY HOOKE. *Robert Hooke was one of the first observers to study the Moon in detail. This drawing shows some craters in recognizable form, and is remarkably accurate considering the low-powered telescope with which it was made*

Hooke admitted that the new reflecting telescope was effective, and hinted that he had himself made one as long ago as 1664. However, he was strongly critical of Newton's theories about light, and Newton was a man who hated criticism. He was touchy and sensitive, and was always reluctant to become involved in arguments. He did have some angry exchanges with Hooke and others, but on the whole he preferred to say as little as possible, and he even refused to publish parts of his scientific work. Indeed his most important studies on the nature of light, contained in his book *Opticks*, remained unpublished until 1704, after Hooke's death.

Other leading members of the Royal Society at that period were Edmond Halley, who later became famous because of his work concerning the comet which now bears his name, and Sir Christopher Wren. Wren is best remembered as the great architect who was responsible for the design of St. Paul's Cathedral following the Great Fire of London, but he was also an astronomer, and had at one time been a Professor of Astronomy at Oxford University. In 1684 Hooke, Halley and Wren discussed the problem of gravitation, and came to the conclusion that what we now term the *inverse square law* must be true.

Hooke had published a book on gravitation, containing many of the same conclusions as those which Newton had reached during his years at Woolsthorpe. Hooke, then, knew that the force between any two bodies will become weaker if the bodies are moved farther apart, and he believed that he had discovered the amount of this weakening. It can be explained by simple arithmetic, so let us take a convenient case of two planets which revolve round the Sun at distances of 2 million miles and 5 million miles respectively, as shown in the diagram on page 75. (No planet is so close as this; Mercury, the nearest-in, is 36 million miles from the Sun, but the basic theory is just the same.)

HYGROMETER MADE BY HOOKE. *This was one of the many scientific instruments developed by Hooke; indeed, his researches extended into almost all branches of science*

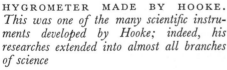

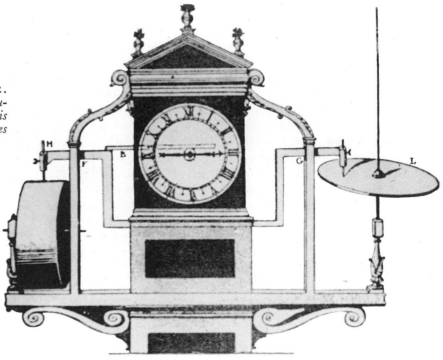

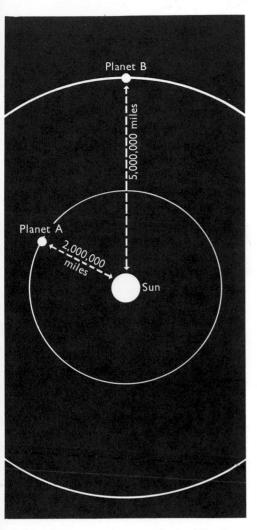

THE INVERSE SQUARE LAW. *The diagram illustrates the case of two hypothetical planets, one at 2,000,000 miles from the Sun and the other at 5,000,000 miles. By application of the inverse square law, Newton discovered the relative gravitational force exerted on the two bodies. Of course, no planets exist so close to the Sun—Mercury, the innermost member of the Solar System, has an average distance of 36,000,000 miles—but the principles are exactly the same whatever the distances may be. Various scientists, including Hooke, had realized that the inverse square must hold good, but only Newton could at that time produce a valid mathematical proof*

Two squared, or 2×2, is 4. Five squared, or 5×5, is 25. Then according to the inverse square law, the Sun's force on the two planets will be in the ratio of $\frac{1}{4}$ to $\frac{1}{25}$, so that the force on the more distant planet will be only $\frac{4}{25}$ of that on the nearer planet. If this is so, then it can be shown that each planet will move in an orbit which is not a circle, but an ellipse.

Hooke guessed this, but he was not a good enough mathematician to prove it. Neither were Halley and Wren, and eventually Halley went down to Cambridge to consult Newton. He was very surprised to learn that Newton had solved the problem years before, but had not announced it, and had even lost his notes!

Halley persuaded Newton to rework his calculations and to allow them to be published. Newton agreed, and during 1685 and 1686 he worked on the book which he called the *Philosophiæ Naturalis Principia Mathematica* ('Mathematical Principles of Natural Philosophy'), but which is known to everyone simply as the *Principia*.

Halley's idea was that the book should be published by the Royal Society, but there were money difficulties; the Society had just issued a tremendous book by Francis Willughby called *The History of Fishes*, and the book had been a financial failure. (Later, when Halley was a salaried official of the Royal Society, he was presented with fifty copies of *The History of Fishes* instead of being given £50 which the Society owed him. Halley is known to have had a strong sense of humor, but whether he was amused or not remains uncertain.)

There was the added trouble that Newton was being constantly irritated by his disputes with Hooke and others, and he was quite capable of changing his mind and withdrawing the *Principia* altogether. Halley knew this, and generously offered to pay for the publication out of his own pocket. This was done, and the book appeared in print.

The *Principia* had taken Newton fifteen months to write, and has been described as the greatest mental effort ever made by one man; it has ensured that Newton's name will live for all time. Many of the age-old problems of astronomy were solved at one stroke. As well as dealing with gravitation and the movements of the planets, the *Principia* contained sections dealing with matters such as the tides, whose cause had not previously been understood. It is impossible to describe the book in a matter of a few lines, but those who have studied it can only wonder that a man still aged less than fifty could have accomplished so much.

Newton lived for forty years after the publication of the *Principia*, and continued his scientific work. He was also active in other directions; he entered Parliament for a time, and became Master of the Royal Mint, helping to revise Britain's coinage—which badly needed attention, since many of the coins in circulation had been reduced in value by having pieces chopped off them. In 1705 he was knighted by Queen Anne, and he also became President of the Royal Society. Two further editions of the *Principia* were issued before Newton died in 1727, and was—fittingly—buried in Westminster Abbey.

Newton was not infallible. He made mistakes, and some of his ideas sound strange today. For instance he spent a great deal of

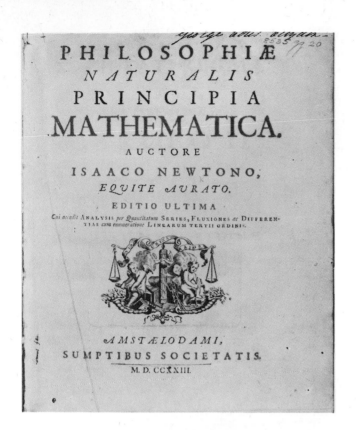

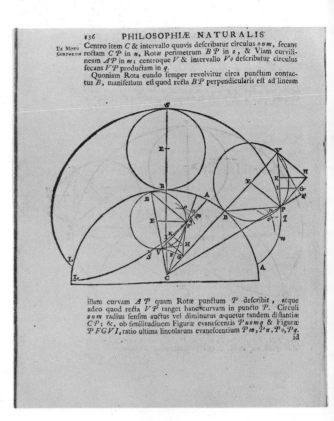

TITLE-PAGE AND SPECIMEN PAGE FROM THE *Principia*—or, *to give the book its full title* Philosophiæ Naturalis Principia Mathematica. *This has often been regarded as the greatest scientific work ever produced*

time experimenting in 'alchemy', the so-called science of making gold from other materials. It would be wrong to suppose that gold-making was the only aim of the alchemists; they were searching for hidden truths, and they also hoped to find the key to ever-lasting life and eternal youth. Newton regarded this as quite possible, and he also made careful studies of old writings in an effort to find hidden meanings in them; he left a mass of notes and manuscripts which are only now being properly studied to see whether they have any value except to the historian.

Newton was touchy and intolerant, and this was one reason why he became involved in so many disputes. For instance there was a long argument with a German mathematician, Gottfried Leibnitz, as to who had invented the calculus. The truth of the matter is that Newton's 'method of fluxions' and Leibnitz' calculus had been developed at about the same time; very probably Newton was the first to make use of it, but Leibnitz' methods were more convenient, and became universally adopted. The quarrel was long and bitter, and was still raging when Leibnitz died in 1716.

Yet Newton's scientific errors were relatively few, and his *Principia* did more for astronomy than any other book before or since. Those who follow its arguments, and who bear in mind that it represented only a part of Newton's work, will hardly question that he was the greatest scientific genius not only of his own time, but perhaps of all time.

13 the royal observatory

HERSTMONCEUX CASTLE, *site of the present-day Royal Greenwich Observatory. The Castle, which is near Hailsham in Sussex, was photographed by Patrick Moore while alterations to it were still being made*

OLD GREENWICH. *The original Royal Observatory as it must have appeared in Flamsteed's time*

THERE ARE STILL some people who regard astronomy as a useless science—interesting, no doubt, but of no practical value. Such people forget that astronomy is the basis of all timekeeping and navigation, as well as being of practical use in other directions. It is true, for instance, that Britain's leading observatory, Greenwich, was founded in 1675 at the express order of King Charles II to assist British sailors.

Britain has always been a seafaring nation, but in the seventeenth century maps were by no means accurate. To make matters worse, navigation was still in the hit-or-miss stage. After a voyage lasting for a week or two, sailors far out of sight of land seldom had much idea of where they were.

To fix one's position on the surface of the Earth it is necessary to obtain latitude and longitude. Finding latitude presents no particular difficulty, since it may be obtained by measuring the apparent positions of the stars; as we have seen, the altitudes of the Celestial Pole above the horizon, given in degrees, minutes and seconds of arc, is equal to the latitude of the observer. In the northern hemisphere of the Earth, therefore, all that is necessary is to measure the height of the Pole Star and then make a slight correction to allow for the fact that the Pole Star is not exactly at the polar point. There is no bright south polar star, but a sailor of the Middle Ages could always work out his latitude with reasonable accuracy. The real problem lay in finding the longitude.

A ship's longitude is the difference between the meridian which

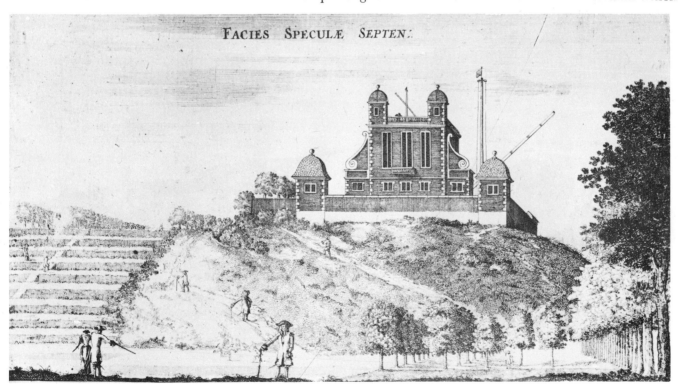

FACIES SPECULÆ SEPTEN:

78 **the royal observatory**

THE ROYAL OBSERVATORY IN FLAM
STEED'S TIME. *An interior view, showing
observers with quadrant and telescope*

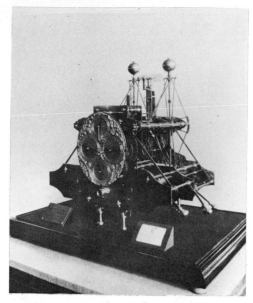

HARRISON'S NO. I MARINE CHRONO-
METER. *This was the first 'clock' which
proved capable of keeping time accurately over
long sea voyages*

THE MERIDIAN LINE AT GREENWICH
OBSERVATORY. *This line at the old
Royal Observatory (now known as Flamsteed
House) marks the boundary between the
eastern and western hemispheres. Reckon-
ing has not been altered by the shift of the
Royal Observatory from Greenwich Park to
Herstmonceux. Photograph by Patrick Moore*

she happens to be on, and a standard meridian such as that of
Greenwich. Local noon is easily found, since this is the moment
when the Sun is at its greatest height above the horizon. A reliable
clock will give Greenwich time at this particular moment, and so
the longitude of the ship may be calculated.

Unfortunately sailors of Charles II's day had no good clocks.
Christiaan Huygens had attempted to produce a timekeeper
which would be useful to sailors, but was unable to make it accurate enough, and it was not until the following century that John
Harrison, son of a Yorkshire carpenter, developed a *chronometer*
which was adequate for use at sea.

As long ago as 1474 Regiomontanus had suggested that the best
way to measure longitude would be to determine the position of
the Moon with respect to the stars. Since the Moon is so close to us
on the astronomical scale, it moves across the starry background
at a rate of about 13 degrees per day. The position of the Moon
among the stars therefore acts as a clock to indicate the time; and
once the time is known, longitude can be worked out. The need,
then, was to measure the Moon's position accurately. It was also
necessary to have a really reliable star catalogue.

The best catalogue available was Tycho Brahe's. However, the
star positions had been drawn up with the aid of instruments
without telescopic sights, and a better catalogue had become
urgently needed.

A Frenchman visiting England, Saint-Pierre, put forward an
alternative scheme which also involved knowing the positions of
the Moon and stars. His method, unlike that of Regiomontanus,
was not practicable, but the King heard about it, and appointed a
committee of Royal Society members to investigate the whole
problem of longitude-finding. The committee was headed by the
Rev. John Flamsteed, already known as a skilful astronomer. They
presented a report, saying that longitudes could indeed be found
by using the 'Moon clock'—if only there were a good enough star
catalogue. Charles accordingly ordered that a special observatory
should be set up, and a new star catalogue produced for the use of
British seamen.

The site selected was the Royal Park at Greenwich, which was
then a small village well outside London. Typically, the King
raised the money for it by the sale of 'old and decayed' gunpowder, and the original buildings—which still stand—were
designed by Sir Christopher Wren. Flamsteed was appointed
royal astronomer, later given the official title of Astronomer
Royal, and instructed to begin work on the catalogue. However,
the King's generosity did not extend to supplying telescopes or
other instruments; Flamsteed was expected to provide these for
himself!

Flamsteed was born at Denby, near Derby, in 1646, and took
his degree at Cambridge. He was never strong, and he was a
sensitive, irritable man who quarrelled not only with Newton,
but also with many other leading men of the time. Yet it is difficult
to blame him. He began his great work under tremendous
difficulties, and his salary was so small that he had to earn extra
money as well as carrying out his work as Astronomer Royal.
He was a brilliant observer, and as time passed by he was

PROSPECTVS INTRA CAMERAM STELLATAM.

FLAMSTEED HOUSE—*the old Royal Observatory as it is today, showing the Octagon Room surmounted by the time-ball. Photograph by Patrick Moore, 1961*

able to add to the observatory equipment as well as engaging assistants.

Trouble began when Newton asked for his observational results, and Flamsteed was reluctant to provide them—mainly because he was not completely satisfied with them. The quarrel was patched up for a while, and in April 1704 there is a record that Flamsteed and Newton met at Greenwich, apparently on friendly terms. At this meeting Newton asked for a report on the star catalogue for which astronomers all over the world were waiting. Flamsteed replied that he was almost ready for printing arrangements to be made, and Prince George of Denmark, husband of the new sovereign, Queen Anne, generously promised to provide the necessary money for publication.

Still Flamsteed was not quite ready, but he handed the Royal Society committee a copy of his observations as well as an incomplete manuscript of the catalogue itself. He made it clear that the catalogue was not to be printed as it stood, but was to wait until it had been completed and checked; the observations, however, could be produced, and printing was begun.

Four more years went by, and still Flamsteed did not submit his finished catalogue. Other disputes arose, and came to a head in 1711 with the publication of Flamsteed's observations. They

HALLEY'S COMET AND VENUS,
photographed by Slipher in 1910

took the form of a large book containing not only the observations which Flamsteed had passed for publication, but also the star catalogue, which he had not. What had happened was that the Royal Society committee had become tired of waiting, and had asked Edmond Halley to make the best of things. Halley had therefore supplied whole pages of material on his own account, and had added a preface which could not be anything but harmful to Flamsteed's reputation.

Flamsteed was angry and indignant, particularly with Halley and Newton. He wanted to revise the catalogue and reissue it, but Newton held some of the observations and refused to give them up. In 1715 a large number of copies of the book fell into the Astronomer Royal's hands—and he publicly burned them 'that none might remain to show the ingratitude of two of his countrymen'. Flamsteed himself died in 1719, but the revision of the catalogue was finished by two of his assistants, Crosthwait and Sharp, and published in 1725.

At least the final catalogue, published by Crosthwait and Sharp under the title *Historia Cælestis*, proved to be well worth waiting for. It included nearly 3,000 stars, and was far more accurate than Tycho's. It represented the first major contribution to science provided by the Royal Observatory, and it ensured that Flamsteed will always be remembered as one of the greatest of astronomical observers.

However, it did not solve the problem of longitude-finding. As well as knowing the positions of the stars, one has to have a sound knowledge of the movements of the Moon. Flamsteed had not been able to pay a great deal of attention to the Moon, and this question was tackled by his successor as Astronomer Royal, Edmond Halley.

Halley was born in 1656. His parents were well off, and he did not have to face money troubles. He went to Oxford, but left before taking his degree in order to sail for the island of St. Helena, where Napoleon Bonaparte was exiled a century and a half later. There was a good reason for Halley's journey. Tycho had catalogued the northern stars, and Flamsteed was just beginning work on the Greenwich catalogue, but the southern stars which never rise over Europe had been completely neglected. Halley's main object was to catalogue them as well.

St. Helena was not a particularly good choice, and Halley was troubled by bad weather. Nevertheless he was able to draw up a catalogue of 381 stellar positions, and well earned his nickname of 'the Southern Tycho'. When he returned and published his observations he became world famous, and Oxford University granted him an honorary degree.

Halley first came into close contact with Newton at the time of the 'inverse square law' discussions, and the two men always remained on friendly terms. As we have seen, Halley was mainly responsible for the appearance of the *Principia*, and made himself responsible for the cost of printing. This may well be regarded as his greatest contribution to science; without Halley, Newton's immortal book might never have seen the light of day.

Just as Newton's name is always linked with gravitation, so Halley's is associated with the famous comet. In fact it was

HALLEY'S MAGNETIC CHART. *Though astronomy was Halley's main study he also paid attention to other branches of science. During his voyage to St. Helena he noticed that the compass needle did not point due north, and from this he deduced—correctly— that the north magnetic pole is not situated exactly at the geographical pole. Years later, in 1698, he was given a commission in the Royal Navy, and sailed as captain of a ship, the* Paramour, *on a journey which enabled him to study the 'variation' of the compass— that is to say the difference between true north and magnetic north. In a second cruise, in 1699–1701, he went to the South Atlantic and encountered Antarctic icebergs. A year later he was able to publish a chart showing the magnetic variations over the whole world. Sea navigators found it remarkably useful*

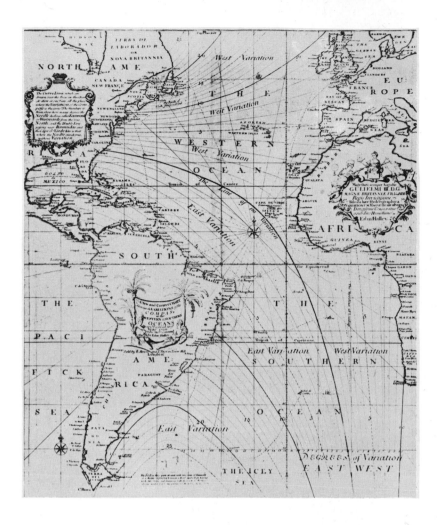

HALLEY'S CHART OF THE SOUTHERN SKY

Halley's book about comets, published in 1705, which provided the most dramatic proof that the laws given in the *Principia* were correct.

A brilliant comet with a tail stretching half-way or more across the sky is a glorious spectacle. Not unnaturally the ancients found it frightening, and an indication that the gods were angry, so that disasters of all kinds were liable to follow. For instance the Roman writer Pliny, who was killed during the eruption of Vesuvius in A.D. 79 which overwhelmed the towns of Pompeii and Herculaneum, once wrote that 'we have in the war between Cæsar and Pompey an example of the terrible effects which follow the apparition of a comet . . . that fearful star which overthrows the powers of the Earth, showing its terrible locks'. A bright comet was also seen just before the Battle of Hastings in 1066, and is shown in the famous Bayeux Tapestry which is said (perhaps wrongly) to have been woven by William the Conqueror's wife. Even in Halley's time the old fears lingered on, and in backward countries they are not quite dead yet.

The real mystery was that nobody knew what comets were, or how they moved. They could not be predicted; they would appear without warning, and remain striking for a few days, weeks or months before fading gradually away. They did not seem to flash across the sky in the manner of shooting-stars, but their shifts

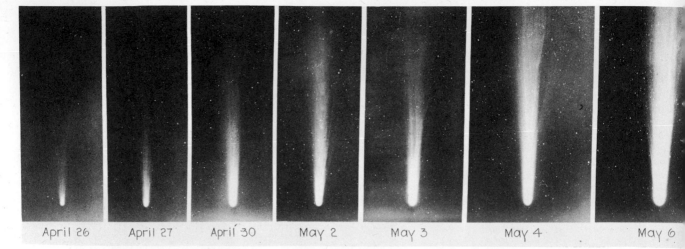

HALLEY'S COMET IN 1910, *photographed at Mount Wilson on various dates*

April 26   April 27   April 30   May 2   May 3   May 4   May 6

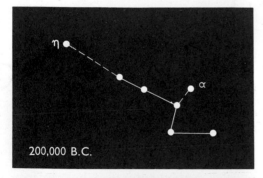

200,000 B.C.

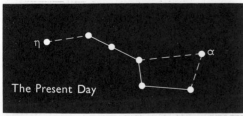

The Present Day

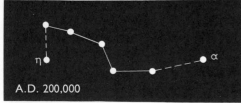

A.D. 200,000

PLOUGH OR DIPPER. *The seven stars in the past, present and future. Two of the stars, α (Dubhe) and η (Alkaid) are moving in a different direction from the other five*

against the starry background were very noticeable from night to night.

Halley was extremely interested in the matter, and he collected all the recorded observations of comets which had been seen between the years 1337 and 1698. One of these comets was that of 1682, which he had himself observed. He calculated the various orbits, using Newton's principles, and realized that the three comets of 1531, 1607 and 1682 had moved in almost identical paths. Could it be that these were merely returns of one and the same comet, moving round the Sun in a period of about seventy-six years? Halley believed so, and he predicted that the comet would again be seen in 1758. By that time he knew that he would be dead, but, modestly, he added: 'If the comet should return according to our prediction, about the year 1758, impartial posterity will not refuse to acknowledge that this was first discovered by an Englishman.'

His forecast came true. On Christmas Day 1758 a German amateur astronomer named Johann Palitzsch, living on his farm near Dresden, rediscovered the comet, and throughout the early part of 1759 it made a brave show in the sky. Fittingly enough, it became known as Halley's Comet; it came back once more in 1835 and 1910, and is due again in 1986.

A comet is not a solid, rocky body similar to a planet. It is made up of numerous small particles surrounded by an 'envelope' of thin gas, and is not nearly so important as it may look; it is of very small mass, and is quite harmless—in fact the Earth passed through the tail of Halley's Comet in 1910 without being in the least damaged. Nowadays it is known that faint comets, visible only with the help of telescopes, are very common, though few really brilliant comets have appeared since 1910.

Comets may be divided into two types. There are the so-called periodical comets, which move round the Sun in elliptical paths; some of these have periods of only a few years, but Halley's is the only bright periodical comet with a period of less than five centuries. The remaining 'great' comets, which attract general attention, move in much more eccentric orbits, so that they come close to the Sun and the Earth only at long intervals—a thousand, ten thousand or a hundred thousand years, perhaps. This is why we cannot tell when to expect them. A comet is not basically self-luminous (though it does emit a certain amount of light when

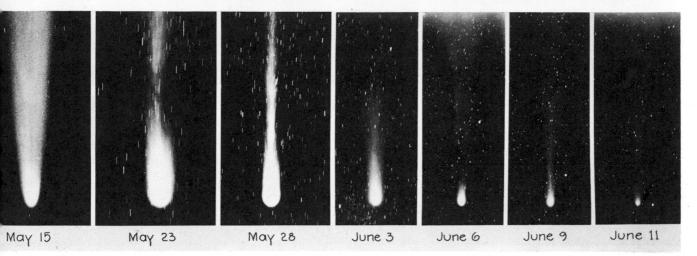

May 15   May 23   May 28   June 3   June 6   June 9   June 11

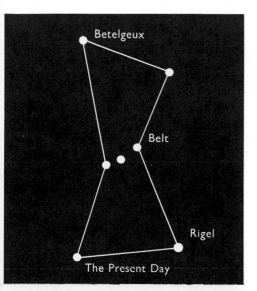

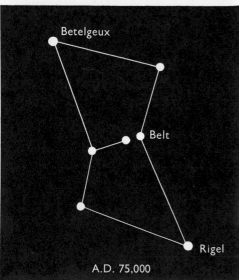

PROPER MOTIONS OF STARS IN ORION. (Upper) *Orion at the present time.* (Lower) *the constellation as it will be in* A.D. *75,000*

close to the Sun), and can be seen only when it is reasonably near perihelion.

Halley was the first to explain the way in which comets move, and he provided a splendid confirmation of Newton's theories. We now know that Halley's Comet was the object which so alarmed the Saxons in 1066, and earlier returns have been traced back to well before Julius Cæsar's time.

It is worth noting that many of the faint comets, too, move in very eccentric paths, so that they have immensely long periods. Such were the comets of spring and autumn 1957 and spring 1970, all of which became bright enough to be seen without a telescope.

Another great discovery made by Halley was that three bright stars, Sirius, Procyon and Arcturus, had shifted slightly since the time when Hipparchus had drawn up his star catalogue. Sirius had indeed shown definite motion even since Tycho Brahe's work at Hven. This was the first indication of *proper motion*, and showed that the old term of 'fixed stars' was misleading.

Flamsteed died in 1719, and Halley was the obvious choice to succeed him as Astronomer Royal. Unfortunately Flamsteed had bought all the instruments himself, and on his death his widow removed them, so that Halley also was left with an observatory but no equipment. He had to begin again, and collect new telescopes and measuring instruments. This he did, and some of his equipment is still to be seen in the 'Octagon Room' which was designed by Sir Christopher Wren.

By this time Halley was well over sixty years of age, but with his usual energy he returned to the problem of longitude-finding. Flamsteed had produced the star catalogue; Halley set out to study the movements of the Moon. It took him nearly nineteen years, but he completed the observations, and they proved to be of immense value—though the 'lunar distance' method of longitude-finding was never used at sea, since the development of the chronometer, by Harrison, made it unnecessary.

Halley made many other important observations. He watched a transit of Mercury during his stay in St. Helena, and worked out a method of using transits to measure the distance of the Sun; he observed a total solar eclipse, and recorded that part of the Sun's atmosphere which we now call the *chromosphere*; and altogether he was responsible for tremendous advances in astronomy. When he died in 1742, to be succeeded as Astronomer Royal by the Rev. James Bradley, the whole scientific world mourned his loss.

14 Lomonosov

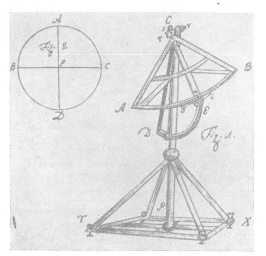

QUADRANT, *designed and used by M. V. Lomonosov*

THE CHAMBER OF CURIOSITIES IN LENINGRAD. *There used to be an observatory at the top of the building, used by Lomonosov. Photograph by Patrick Moore, 1960*

UP TO NOW WE HAVE been describing the lives and researches of astronomers of Western Europe—Britons, Italians, Germans, Frenchmen and the rest. American science had hardly begun, and in the Far East there had been no developments since the ancient Chinese had recorded their comets and eclipses. What, then, of Russia?

Scientists of the modern U.S.S.R. play a leading part in astronomical research, but during the eighteenth century Russia was relatively backward. Even the Copernican theory was generally rejected, both on scientific and on religious grounds, and as late as 1740 most Russian professors held to the old idea that the Sun must go round the Earth. The change-over to a more enlightened view was due in part, at least, to the first of Russia's great astronomers, Mikhail Vasilevich Lomonosov.

Lomonosov was born on an island not far from the city of Archangel in 1711. His father, a free peasant, was a fisherman, and during his boyhood Mikhail went on several fishing expeditions into the White Sea, well inside the Arctic Circle. Like most Russian children of the time he had very little schooling, but by the age of fifteen he had learned how to read and write, and—according to a story which may well be true—he ran away from home and joined a train of sleds carrying frozen fish, bound for Moscow. He enrolled in a school attached to a large monastery, and his real education began. He became interested in science, and also started to write poetry.

He made rapid progress, and came to the notice of the St. Petersburg Academy of Sciences, which had been founded by the Czar Peter I. In 1735 he was sent to the University of St. Petersburg, and a year later he and other students were sent to a German University at Marburg principally to study chemistry and mineralogy.

Lomonosov was already showing signs of exceptional ability. It cannot be said that he behaved well; he shocked his German professors by his habit of drinking too much (a not uncommon failing among Russian students in those days), and when he went back to his own country in 1741 he abandoned the wife whom he had married after his arrival in Marburg. Two years later he insulted some of his colleagues at the St. Petersburg Academy; complaints followed, and he was imprisoned for several months—during which time he wrote two of his most famous poems. However, he became a full member of the Academy in 1745, and for some time was director of the main chemical laboratory, which he himself designed.

Lomonosov's activities were widespread. He was interested in problems in navigation, and with the aid of his own measuring instruments he determined the latitudes and longitudes of many of the main cities of his country, so drawing up the first accurate map of the Russian Empire. He described a 'solar furnace', using lenses and mirrors; he invented a new kind of reflecting telescope,

THE AURORA BOREALIS. *Lomonosov was among those who studied the spectacular auroræ or Polar Lights*

on a pattern not unlike that developed later by Sir William Herschel; he investigated electrical phenomena, including the spectacular auroræ or Polar Lights. All this time he never ceased to·champion the Copernican theory, though many of his colleagues disapproved.

One of Lomonosov's most interesting discoveries was made in 1761, when Venus passed in transit across the face of the Sun for the first time since Horrocks and Crabtree had made their famous observations in 1639.

When Edmond Halley had been at St. Helena, he had watched a transit—not of Venus, but of Mercury. Following up an earlier suggestion by James Gregory, he had realized that such transits might provide a method of measuring the distance of the Sun. Of course there are many refinements to be taken into account, and the method used was somewhat complex, but in theory it was sound enough. Mercury was difficult to measure accurately, but Venus, which is much larger and closer, held out more promise, and Halley certainly regretted that he could not hope to live long enough to watch the 1761 transit.

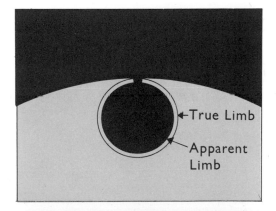

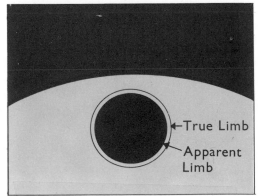

THE BLACK DROP. *This was the effect which ruined the accuracy of the transit of Venus method of determining the Sun's distance*

When the time came the transit was carefully studied by astronomers all over the world. Unfortunately the results were badly affected by a phenomenon which became known as the 'Black Drop'. As Venus passed on to the Sun it appeared to draw a strip of darkness after it, and when this disappeared the transit was already in progress. It was therefore hard to time the moment when the transit actually started—and since this was essential to the method, the distance measurements were not reliable. Similar troubles were experienced at the next transit, that of 1769. Many years later the German astronomer Encke published a final result based on the observations, and gave the Sun's distance as 95,279,000 miles, which is now known to be over two million miles too great.

No reference to transits would be complete without a reference to the amazing misfortunes of the French astronomer Guillaume Legentil, who decided to observe the 1761 transit from

OCCULTATION OF REGULUS BY
VENUS, JULY 7, 1959. *It is very
seldom that Venus passes in front of, and
occults, a bright star, but such an event took
place on the early afternoon of July 7, 1959,
when Regulus in Leo was occulted. These
drawings, made by Patrick Moore with the
12½-inch reflector at Henry Brinton's observatory in Selsey, show the changing position of
Venus and Regulus. For a brief period before
immersion Regulus was shining through the
atmosphere surrounding Venus, and appreciable fading was recorded, so yielding information as to the height of the atmosphere of
Venus. It will be many centuries before Venus
again passes in front of a 1st-magnitude star*

Pondicherry in India. He sailed in a French frigate, but unhappily for him a war between England and France was in progress, and about this time Pondicherry was captured by the English, so that Legentil had to turn back. Before he could reach land the transit was over, and all he could do was to make rough notes from the deck of his ship. Rather than risk a second delay he elected to wait in India for the next eight years, and observe the 1769 transit instead. Again he was unlucky, since clouds covered the Sun at the critical moment. Since the next transit was not due until 1874, Legentil set off for home; twice he was shipwrecked, and reached Paris after a total absence of eleven years to find that he had been presumed dead and that his heirs were just about to distribute his property!

Since those far-off days two more transits of Venus have taken place—in 1874 and 1882—and were closely studied. Once again, however, the results were not as reliable as had been hoped, and it must be admitted that the whole transit method has been found to be unsatisfactory. Future transits will not be regarded as of much importance.

However, it is interesting to note that in this connection Venus has recently been used in a different way. Since 1960, astronomers in Britain, America and Russia have been able to 'bounce' radar pulses off the planet, after which they measure the time-lag between the transmission and the resulting echo. Radar pulses, of

TRANSITS OF VENUS IN 1874 AND
1882. *The diagram shows the apparent
track of Venus across the face of the Sun
during these two transits. In neither case was
the transit central, but it had been hoped to
obtain measures which would be accurate
enough to give a reliable estimate of the Sun's
distance from the Earth. Great attention was
paid to observing the transits, but the results
proved to be very disappointing, and the next
two transits (those of 2004 and 2012) will not
be regarded as of much astronomical importance*

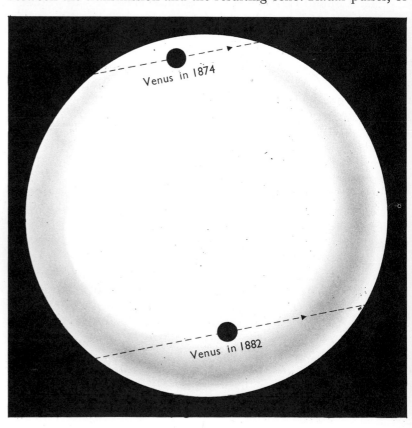

Venus in 1874

Venus in 1882

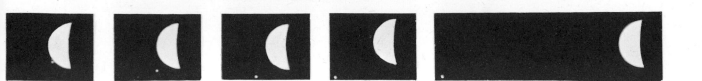

course, move at the same speed as visual light (186,000 miles per second), and so it is possible to measure the distances which the pulses have travelled. The distance of Venus can then be obtained, and this leads, in turn, to the accurate determination of the *astronomical unit* or Earth–Sun distance. The latest value is 92,956,000 miles.

Mikhail Lomonosov watched the 1761 transit from his home in St. Petersburg, where he had erected a refracting telescope of focal length $4\frac{1}{2}$ feet. One thing interested him particularly. Just before the transit began, the Sun's limb seemed to become 'smudgy', and a similar appearance was seen immediately after the transit was over. Moreover, there seemed to be a curious sort of 'blister' just before Venus passed right on to the Sun.

Lomonosov could find only one way to explain what he had seen. In his own words: 'The planet Venus is surrounded by a considerable atmosphere, equal to, if not greater than, that which envelops our earthly sphere.' This observation strengthened Lomonosov's view that some of the planets at least are not so very unlike the Earth, and he even suggested that life might exist on them.

An atmosphere around Venus would indeed account for the appearance, and we now know that Lomonosov was right. Venus has a very deep, dense atmosphere, with a ground pressure of at least 90 times that of our air at sea-level, and made up almost entirely of carbon dioxide. Yet Lomonosov was wrong in supposing that Venus might very well be habitable. Recent results with space-probes have shown that it is far too hot for any highly-developed Earth-type life to exist there. The surface temperature has been measured at about 900 degrees Fahrenheit.

It will be some time before the next transit takes place; one will occur in 2004 and another in 2012, after which we will have to wait until 2117 and 2125. But though past observations have proved something of a disappointment to astronomers, it is interesting to remember that it was at the 1761 transit that Mikhail Lomonosov discovered that Venus, like the Earth, is surrounded by an atmospheric mantle.

VIEW OF RUSSIA'S OLDEST OBSERVATORY. *The Chamber of Curiosities in Leningrad, where the first observatory in Russia was set up and used by M. V. Lomonosov. The photograph was taken from the opposite bank of the river. There is no longer an observatory in the building, and the building itself is now used as a museum. Lomonosov's observations of the 1761 transit of Venus were not actually carried out from here, but from another site in St. Petersburg (now Leningrad). Photograph by Patrick Moore, 1960*

15   the 'father of stellar astronomy'

SIR WILLIAM HERSCHEL, *often regarded as the greatest observer in astronomical history*

OBSERVATORY HOUSE, SLOUGH. *The house where three Herschels—William, Caroline and John—lived and worked. The great 40-foot reflector was erected in the garden. Eventually the house fell into disrepair, and was pulled down in 1960. This photograph taken by Patrick Moore was probably the last picture ever taken of it— three days before demolition was begun*

THE WORK OF NEWTON completed the greatest of all the revolutions in astronomical thought—begun by Copernicus, continued by Galileo and Kepler, and ended with the publication of the immortal *Principia*.

Eleven years after Newton's death, a boy named Wilhelm Herschel was born in Hanover, which was then part of the British Empire. (It remained so, incidentally, until Queen Victoria came to the throne in 1837.) Herschel was to become as famous in his own day as Newton had been earlier, and it is he who has justly earned the nickname of 'the father of stellar astronomy'. As a mathematician and theorist he could not compare with Newton— nor did he wish to. He was an observer first and foremost, perhaps the greatest who has ever lived.

It would be unfair not to mention at least some of the other leading astronomers of the eighteenth century. There were many of them. There was Alexis Clairaut of France, born in 1713, who studied higher mathematics at the age of eleven and sent a valuable paper to the Paris Academy of Sciences when only thirteen; he revised Halley's work on the famous comet, and predicted that it would come back to perihelion not in 1758, but in 1759. Clairaut's work was so accurate that he was in error by less than a month. There were the brilliant mathematicians Leonhard Euler, who continued his work even after he had become blind, and Joseph Lagrange. There was Jean Sylvan Bailly, who was guillotined during the French Revolution in 1793; it was said that 'the Republic has no need of wise men'. And there were many more; but Herschel towered head and shoulders above them when it came to purely observational astronomy.

Friedrich Wilhelm Herschel, better known to us as William Herschel, was born on November 15, 1738. His father was bandmaster of the Hanoverian Guard, and most of his children inherited musical ability. William was no exception, and at the age of fifteen entered the band as oboist.

He soon found that he did not care for Army life. Moreover the Seven Years War was raging, and after some unpleasant experiences during the Battle of Hastenbeck in 1757 he left the band and came to England. It is said that when he landed at Dover he had only one French crown piece in his pocket.

Fortunately he was a gifted musician, and had little trouble in earning his living. In 1765 he became organist at Halifax—after an audition in which he proved to be a better player than his chief rival, a Dr. Wainwright, who was known to be an expert. During the following year he moved to Bath as oboist in an orchestra which played daily in the still-famous Pump Room, and he then became organist at the new Octagon Chapel.

Though he was a professional musician, music was never everything to Herschel. He had always been interested in astronomy, and about 1771 he turned to it as a serious hobby. He hired a small reflecting telescope, and enjoyed using it; but it was not

URANUS, *showing the planet together with one of its satellites. A painting made by D. A. Hardy from an observation by Patrick Moore on January 19, 1954, 23.40 with a power of 500 on a 12.5-inch reflector*

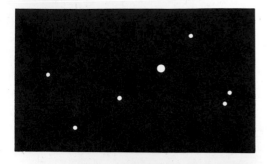

MOVEMENT OF URANUS AMONG THE STARS. *Observations made by Patrick Moore on March 4 and 6, 1960*

6½-INCH REFLECTOR MADE BY WILLIAM HERSCHEL. *This is a replica of the telescope with which Herschel discovered Uranus in 1781*

powerful enough to satisfy him, and he looked round for something larger. Unfortunately he found that the price of a bigger telescope was more than he could afford. The only solution was to make his own.

To be of any use at all, a telescope mirror must be very accurate indeed. If the curve is at all irregular the instrument will produce nothing but blurred images. On Newton's pattern there are two mirrors: the main one or *speculum*, and the smaller *flat*. The most important and expensive part of the instrument is the speculum. Nowadays, mirrors of this kind are made of glass, but during the eighteenth century they were more commonly made of a special alloy known as speculum metal.

Herschel decided to become a telescope-maker, and in June 1773 he began work in earnest. He had to continue his musical profession, but all his spare time was spent with his mirrors. Eventually he produced a tolerable 5-inch speculum, but before this he had had over *two hundred* failures.

By this time he had been joined by two of his family; his brother Alexander, also a musician and moreover a skilful amateur mechanic, and his sister Caroline. Alexander left after a time, but Caroline remained with William, and became an astronomer in her own right—indeed, she discovered six comets.

Herschel's first recorded observation, a sketch of the Orion Nebula, was made on March 4, 1774. His first successful reflector opened up new fields to him; he concentrated on making larger and better telescopes as well as on his musical career. Then, in 1781, came the discovery which was to change the whole course of his life.

Herschel was particularly interested in the way in which the stars are distributed in space. He knew them to be suns, and he knew them to be immensely distant. Years earlier Thomas Wright of Durham, originally a clock-maker's apprentice and then a teacher of mathematics, had written a book in which he suggested that the stars might be arranged in the form of a disk. Wright, who was still living at the start of Herschel's observational career (he did not die until 1785) had never followed up this theory; neither had the German philosopher Immanuel Kant, who had put forward a rather similar idea in 1755.

When Herschel began work he had a completely open mind on the whole subject, but he knew that the only way to find out was to undertake a long series of practical observations. He therefore decided to use his telescopes to 'review the sky', counting the stars in certain selected regions and noting their distribution.

While busy at this task, on March 13, 1781, he came across an object in the constellation of Gemini (the Twins) which did not look in the least like an ordinary star. Instead of being a point of light it showed a distinct disk, rather greenish in hue. Herschel was using a home-made telescope of 7 feet focal length and a 6½-inch mirror, and he knew that the instrument was a good one, so that he was confident of the correctness of his observation. During the following nights he discovered that the curious object seemed to be moving slowly against the starry background. Therefore it must be much closer than the stars.

Herschel, naturally enough, took it to be a comet; indeed his paper to the Royal Society giving details of the discovery is entitled *Account of a Comet*. He was interested, but by no means wildly excited.

Once several observations of a moving body have been secured, the orbit can be worked out. Herschel's 'comet' was studied by the Finnish astronomer Anders Lexell, then a Professor at St. Petersburg. Lexell's findings were, to say the least of it, startling. The object was not a comet at all; it was a new planet, moving round the Sun at a distance much greater than that of Saturn.

Astronomers all over the world were taken completely by surprise. It had not occurred to them that Saturn might not after all be the most remote planet, and the discovery was completely unexpected. Yet there could be no doubt that Lexell's calculations were right.

Herschel became famous at once, and the King, George III, appointed him royal astronomer (not Astronomer Royal; the official post was at that time held by the Rev. Nevil Maskelyne) at a salary of £200 per year. The grant was not enough to make Herschel rich, but it did allow him to abandon music as a career, so that he could spend the rest of his life on astronomical work.

In gratitude to the King, Herschel proposed to name the new planet 'Georgium Sidus'—the Georgian Star. Foreign astronomers naturally objected, and finally the planet was christened Uranus, after one of the mythological gods.

Uranus can just be seen with the naked eye, but it is a very faint object of below the 5th magnitude, and to identify it without the help of a telescope would be practically impossible. However, it had been recorded before Herschel's time. Flamsteed had seen it as long ago as 1690. Pierre Lemonnier, a French Professor who had been born in 1715, had noted it a dozen times between 1750 and 1771, and had he taken the trouble to compare his observations he would certainly have detected its motion among the stars. However, Lemonnier was not a methodical man; it is even stated, apparently with truth, that one of his observations of Uranus was found years later scrawled down on an old paper bag which had contained hair perfume. (It is also related that Lemonnier was so ill-tempered that he quarrelled with almost everyone whom he met.)

Uranus proved to have a mean distance from the Sun of 1,783,000,000 miles, as against the 886,000,000 of Saturn. It is so remote, and moves so slowly, that it takes eighty-four years to complete one journey. It is a large world over 29,000 miles in diameter, so that it could contain about fifty globes the size of the Earth. However, it is by no means Earthlike; it is made up chiefly of gas, and it is intensely cold.

It is probably true to say that Herschel is best remembered as being the discoverer of Uranus, but this gives a wrong idea of his career. His main work was in the study of the stars, and had he not been busy in his systematic reviews of the sky he would not have found his new planet.

Herschel believed that the apparent brightness of a star must be a reliable guide to its distance from us, so that brilliant stars would be closer than fainter ones. This is not the case, but Herschel had no means of working out the distance of any star, and his assumption was perfectly reasonable.

URANUS AND TWO OF ITS SATEL-LITES. *Photograph by B. Warner and T. Saemundsson, University of London Observatory, 1960. Of the five satellites of Uranus, one (Miranda) is excessively faint; two (Ariel and Umbriel) are difficult objects, and the outer two (Titania and Oberon) are easier, though a moderate telescope is needed to show them. In this photograph, Titania and Oberon are shown. Surface detail on Uranus itself is almost impossible to record photographically*

COMPARATIVE SIZES OF URANUS AND THE EARTH. *Uranus, with a diameter of 29,300 miles, is a giant planet, though considerably smaller than either Jupiter or Saturn*

STARFIELD IN ORION *showing the gaseous nebula M42, photographed with a 12-inch reflector by H. R. Hatfield in 1971*

The way to find out how the stars were arranged, therefore, should be to count them. The regions of the sky which contained the greatest number of stars would represent the greatest extensions of the stellar system or *Galaxy*. It was impossible for Herschel to count every star revealed by his telescopes, so he adopted the method of 'star-gauging', or counting the stars in certain carefully-selected regions.

There are obviously more stars in and near the Milky Way than in other areas, but Herschel soon found that the percentage increase was greater for faint stars. For instance suppose that we take two areas, one near the Milky Way and one remote, and use a telescope which will show four times as many bright stars in the Milky Way zone than in the other. With a larger telescope, it will be found that the faint stars are ten times more common than the bright ones. In other words, faint stars are unexpectedly numerous close to the Milky Way, and Herschel came to the conclusion that the Galaxy must be shaped rather like a double-convex lens or two plates clapped together by their rims, with the Sun somewhere near the middle.

The suggested arrangement is shown in the diagram. If S represents the Sun, most stars will be seen in the directions SA and SB, and this will mark the Milky Way—a mere perspective effect. Fewer stars will be seen in directions SC and SD, indicating the most barren parts of the sky in the regions of the constellations of Lynx and Sculptor.

In the main Herschel was correct. His only serious error lay in placing the Sun in the middle of the system; we now know that it is well out to one side. However, this error was not Herschel's fault, and his idea of a lens-shaped Galaxy was a great step forward.

At about the same time he managed to measure the movement of the Sun itself with respect to the stars. This was done by means of a well-known effect. If you drive along a road with trees to either side, you will note that the trees ahead of you seem to 'open out' as you approach them, while the trees behind will 'close up'. Herschel found that the stars were behaving in the same way. The shifts were remarkably small and difficult to measure, but in

HERSCHEL'S PLAN OF THE GALAXY. *Herschel was the first man to propose a plan of the Galaxy or 'Milky Way' which proved to be reasonably accurate, though one or two earlier guesses had been on the right lines. Herschel's main error was in placing the Sun near the centre of the system, as shown here; in reality it is well out toward the edge. If an observer looks along the direction AB he will see many stars in much the same direction, giving rise to the Milky Way effect; if he looks along direction CD he will see fewer stars. In this respect Herschel's views were perfectly sound*

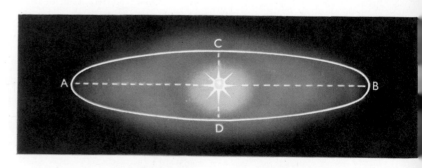

THE GREAT NEBULA IN ORION. *This cloud of gas is visible to the naked eye. It lies in the Sword of Orion, and is 1500 light-years away. The gas is made luminous by the presence of stars contained in the cloud. Even at its densest parts, these clouds are much more rarefied than the best vacuum obtainable on Earth, but due to its enormous size the nebula has a total mass ten times that of our Sun. Photograph taken by the 200-inch Hale reflector at Palomar*

A 6-INCH NEWTONIAN REFLECTOR.
*The tube is a skeleton, which is perfectly
satisfactory for instruments of this type*

GALAXY IN CASSIOPEIA, N.G.C.
(=New General Catalogue) 147. *The
photograph was taken in red light with the
200-inch Hale reflector at Palomar. This is an
external system, and lies far beyond the
boundaries of our own Galaxy*

1783 he announced that the Sun is moving towards a point
in the constellation Hercules, not far from the brilliant Vega.
The position which he gave is extremely close to the modern
estimate.

Meanwhile he had felt the need of a still larger telescope, and
he developed a new pattern for a reflector. In Newton's arrangement the flat lies in the main tube, and so blocks out a little of the
starlight from the main mirror. Moreover there are two reflections
—one from the speculum and one from the flat—so that further
light is lost. Herschel decided to tilt his speculum at an angle, and
bring the light to focus at the upper end of the tube, thus doing
away with the flat altogether. In fact, the arrangement is not so
good as might be thought, and Herschelian telescopes are seldom
made nowadays.

In January 1787 Herschel tested one of his new-type reflectors,
a giant instrument with a focal length of 20 feet. Almost at once he
made the interesting discovery that Uranus, which he had first
identified six years earlier, was attended by two satellites—now
known to us as Titania and Oberon. This brilliant result made
him decide to construct an even larger telescope, with a mirror
48 inches across and a focal length of 40 feet.

The task was by no means easy. The first mirror, tested in

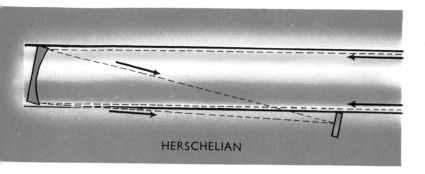

HERSCHELIAN

PRINCIPLE OF THE HERSCHELIAN REFLECTOR. *Herschel's method was to tilt the main mirror, thus doing away with the necessity of having a second mirror or 'flat'. Unfortunately the arrangement has numerous disadvantages, and is now seldom used*

February, was found to be too thin, so that it went out of shape. A second was cast in 1788, but developed a crack and had to be rejected. Finally he made a third mirror, and on August 28, 1789 he used it for the first time. As soon as he looked at Saturn, he saw a new satellite, while another new attendant was detected on September 17. These two newcomers, Mimas and Enceladus, raised the number of Saturnian satellites to seven. As we remember, Huygens had discovered one (Titan) and Cassini four more.

The 40-foot telescope had been set up at Herschel's new home, Observatory House in Slough. The mirror, weighing 2,118 lb., was slung in a ring, and the sheet-iron tube in which it rested was almost 5 feet wide. Ladders 50 feet in length gave access to a movable stage, and the whole instrument was mounted on a revolving platform.

The giant telescope was optically excellent by nineteenth-century standards, but it was clumsy to use, and disturbances in the Earth's air meant that conditions were seldom good enough for it to give of its best. Herschel used it only when the 'seeing' was exceptional, and most of his routine work was done with the smaller 20-foot instrument. The great reflector was used for the last time on January 19, 1811, when it was directed to the Orion Nebula. On New Year's Eve 1839, long after Herschel's death, it was dismounted, and the great tube was laid down on three stone piers in the garden at Slough.

Herschel married in 1788, and his sister Caroline moved out of Observatory House. She remained near at hand, however, living in the town of Slough and coming in every night to help William in his work. Caroline was indeed the most devoted of assistants; she had no wish for personal fame, and was content, as she herself said, with 'reflected glory'. In later life, after William's death, she returned to Hanover, and died in 1848 at the advanced age of ninety-eight.

Herschel's 'reviews of the heavens' resulted in the discovery of many objects. For instance, there were the dim patches known as nebulæ. In 1714 Halley had listed 6; the French astronomer Lacaille compiled a catalogue of 42 in 1755, and Messier, the famous comet-hunter, produced a list of over 100 in the year 1781. Herschel discovered over 1,500 more, and also found that they were of two kinds. Some of them appeared to be made up of stars, while others gave the impression of glowing gas. It was then that Herschel made one of his most inspired suggestions. Could the starry nebulæ be separate galaxies, far beyond the limits of our

HERSCHEL'S 40-FOOT REFLECTOR, *with its scaffolding*

ENGRAVING OF THE 40-FOOT RE-FLECTOR *in the garden of Observatory House, Slough*

FLAMSTEED HOUSE. *The old Royal Observatory in Greenwich Park, now renamed and used as a museum; one room is devoted to Herschel relics. The photograph, taken in 1961 by Patrick Moore, shows the Octagon Room, used by Flamsteed, and the celebrated time-ball*

THE OLD ROYAL OBSERVATORY, GREENWICH. *A view of one of the other buildings. Photograph by Patrick Moore*

Milky Way? We now know that he was right, but the truth of hi suggestion was not proved until over a century later.

Another subject which attracted Herschel was that of double stars, and his studies here led him on to a further important discovery.

A careful look at Mizar, the second star in the 'handle' of the Plough, or Big Dipper—Zeta Ursæ Majoris according to Bayer' nomenclature—shows that it has a smaller star, Alcor, close behind it. On a clear night, anyone with normal eyes can see Alcor with out much trouble, but a telescope shows that Mizar itself is made up of two stars, one *component* being brighter than the other. Mizar is in fact a double.

Pairs of stars are very common. Sometimes, as with Gamma Virginis, the components are exactly equal; sometimes one is much more brilliant than the other—for instance Polaris has a 9th magnitude companion, and other bright stars with dim attendant are Rigel, Antares in the Scorpion, and Sirius.

Up to Herschel's time it had been thought that these double stars were mere effects of perspective. In the diagram, star A i much closer to the Earth than star B; but it happens to lie i almost the same direction, and in a telescope the pair would appear as shown in the inset. Doubles of this sort are known, and are termed *optical doubles*.

Herschel studied double stars closely, and in 1802 he was able t announce that in most cases the two components of a pair reall were close together, revolving round their common centre of

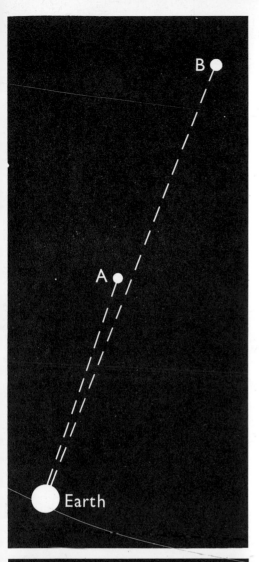

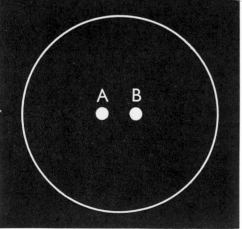

OPTICAL DOUBLE STAR. *As seen from the Earth, stars A and B lie in roughly the same direction, and telescopically they appear as in the circle, though in fact star A is much the closer to us and has no real association with B*

EXTERNAL GALAXY N.G.C.205 (*in Andromeda)—a system far beyond our Milky Way. Herschel was one of the first to suggest that such objects might be external. Photograph taken with the 200-inch Hale reflector*

gravity much as the two bells of a dumbbell will do if twisted by their joining bar. He had in fact discovered real stellar companions, known today as *binary stars*. Rather unexpectedly, it has become clear that binaries are much more common than optical doubles.

These are only a few of Herschel's discoveries. More than any other man, he put stellar astronomy on a really firm footing. Some of his ideas sound strange to us; he was convinced that the Moon and planets were inhabited, and he believed that there might be men living in a cool region inside the Sun. Yet as an observer he was in a class of his own. He was knighted in 1816; he received every honor that the scientific world could bestow, and he became the first President of the newly-formed Astronomical Society of London (now the Royal Astronomical Society). He presented his last scientific paper when he was eighty years old, and he was active almost to the date of his death on August 25, 1822.

His son Sir John Herschel carried on his father's work, and went on an expedition to the Cape of Good Hope, to extend the 'star-gauges' for the southern skies. In all he discovered over 3,300 new double stars and 525 nebulæ. Sir John's own two sons were also active in astronomical work, and Observatory House remained in the possession of the Herschel family for many years.

I last visited it in 1960, when the house was empty and the garden overgrown. The speculum of the 40-foot reflector, which I remember hanging on the wall in the front hall, had been taken away to Flamsteed House, in the old observatory at Greenwich; the plaque marking the site of the great telescope had likewise been removed—but in a shed there still lay part of the tube of Herschel's greatest reflector.

Observatory House was pulled down in the summer of 1960, and has been replaced by a modern block of shops and flats. But though the old building is no longer there, William Herschel himself will never be forgotten.

16   the planet-hunters

A GLANCE AT A CHART of the Solar System will show that the planets are divided into two groups. Mercury, Venus, the Earth and Mars are relatively small and close to the Sun. Then there is a wide gap, followed by the giant planets—Jupiter and the rest. It looked almost as though there were a missing planet between Mars and Jupiter. Kepler suspected that some such body might exist, and even wrote: 'Between Mars and Jupiter I put a planet.'

In 1772 the Director of Berlin Observatory, Johann Elert Bode, drew attention to a strange 'law' concerning planetary distances. It is still known as Bode's Law, though it had been discovered some years earlier by another German, Titius of Wittenberg.

Take the numbers 0, 3, 6, 12, 24, 48, 96 and 192, each of which apart from the first—is double its predecessor. Now add 4 to each, giving: 4, 7, 10, 16, 28, 52, 100 and 196. Taking the Earth's distance from the Sun as 10, these figures give the distances of the remaining planets with remarkable accuracy, as the following table shows:

BODE'S LAW

PLANET	DISTANCE FROM THE SUN	
	ACCORDING TO BODE'S LAW	ACTUAL
Mercury	4	3·9
Venus	7	7·2
Earth	10	10
Mars	16	15.2
—	28	—
Jupiter	52	52·0
Saturn	100	95·4
Uranus	196	191·8

Could this be coincidence? When Bode first made the Law known, Uranus was still undiscovered; but when it was found, nine years later, it fitted excellently into the scheme.

The only trouble was that there appeared to be no planet corresponding to Bode's number 28—the gap between Mars and Jupiter.

If such a planet existed it would certainly be faint, probably invisible without a telescope; if it had been even of the 5th magnitude, astronomers who had worked for years at compiling star catalogues—as Bode himself had done—would certainly have found it. Moreover, it was reasonable to suppose that its orbit would not be far removed from the plane of the ecliptic, so that the planet would lie somewhere in the Zodiac.

In 1800 six astronomers assembled at the German town of Lilienthal, not many miles from Bremen. Johann Schröter, who was famous mainly because of his studies of the Moon, had an observatory there and was a tireless worker in astronomy, so that

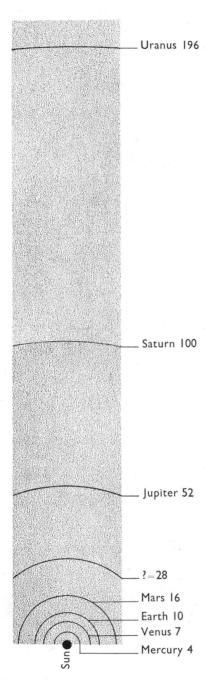

DISTANCES OF THE PLANETS FROM THE SUN ACCORDING TO BODE'S LAW. *Uranus, discovered in 1781, fitted excellently into the scheme. Note the 'gap' corresponding to number 28; it was this which led to an organized search, and the discovery of the first four Minor Planets*

Uranus 196

Saturn 100

Jupiter 52

? = 28

Mars 16
Earth 10
Venus 7
Mercury 4

Sun

PLAN OF SCHRÖTER'S OBSERVA-
TORY AT LILIENTHAL. *This observatory
was for many years the 'astronomical centre'
of Germany and, indeed, of much of Europe.
It was from Lilienthal that the planet-hunt
was organized, and it was here too that
Schröter carried out almost all his pioneer
work in lunar observation*

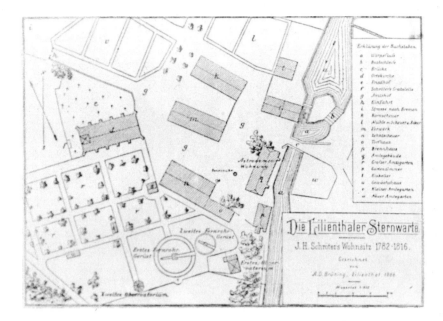

Lilienthal was a good choice. The purpose of the meeting was to organize a systematic hunt for the missing planet. The astronomers formed an association which became known unofficially as the 'celestial police', and each undertook to study a different part of the Zodiac, checking all the stars in an attempt to find an object which moved. The secretary of the 'police' was a Hungarian, Baron Franz Xavier von Zach, who was Director of the Seeberg Observatory near Gotha, and was particularly interested ·in international co-operation among astronomers.

A plan of this sort takes some time to bring into working order, and before the 'police' could begin serious operations they were forestalled. At Palermo, in the island of Sicily, the Italian astronomer Giuseppe Piazzi was busy drawing up a star catalogue, and on January 1, 1801—the first day of the new century—he picked up a point of light which behaved in a most unstarlike manner. It moved appreciably from night to night, and Piazzi thought that it must be a tailless comet. He was interested enough to write to Von Zach, but unfortunately his letter took some time to arrive, and by the time Von Zach received it the strange moving body had been lost in the rays of the Sun.

Luckily Piazzi had made several observations of it. Karl Friedrich Gauss, one of the greatest of German mathematicians, worked out an orbit and forecast its position, so that it was rediscovered exactly a year after Piazzi had first seen it. It did indeed prove to be a new planet, and was named Ceres, in honor of the patron goddess of Sicily.

In one way Ceres seemed to be disappointing. It was too dim to be seen with the naked eye, and was less than 500 miles across, which meant that it was much too small to be regarded as a proper planet. On the other hand its orbit fitted in with Bode's Law, and many astronomers considered that the Solar System was at last complete.

The 'celestial police' had their doubts. It was at least possible, they considered, that other small planets might come to light, and

PATH OF CERES AMONG THE STARS.
*The position of Ceres is shown for May 2, 12
and 22, 1959, close to the star Beta Libræ. At
this time the planet Jupiter was near by, in
the adjacent constellation of Scorpio, the
Scorpion. Drawing by H. P. Wilkins*

INCLINATION OF THE ORBIT OF PALLAS, *which amounts to over 34 degrees*

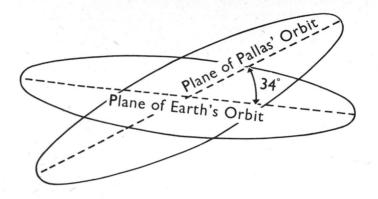

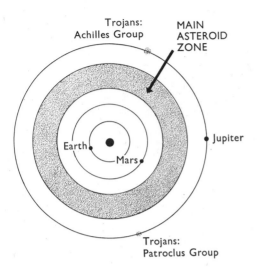

ORBITS OF THE TROJAN ASTEROIDS. *The Trojans revolve round the Sun at mean distances equal to that of Jupiter, so that in effect they move in Jupiter's orbit. One group of Trojans lies well 'ahead' of Jupiter and the other group well 'behind', so that there is no danger of close encounters with Jupiter. The Trojans do not remain precisely 'fixed' in relation to Jupiter, but the diagram suffices to show the general situation*

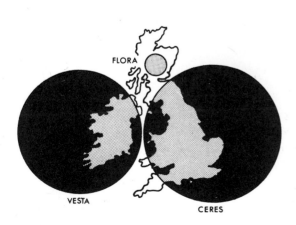

SIZES OF MINOR PLANETS COMPARED WITH THE BRITISH ISLES. *Ceres is the largest of the minor planets, and Vesta the brightest*

they went ahead with their plans. In March 1802 Heinrich Olbers, one of the hunters, detected a second body, slightly smaller than Ceres and moving at a slightly greater distance. Pallas, as it was named, was unusual in one respect; its orbit was inclined to the ecliptic at an angle of over 34 degrees, as against the 10½ degrees of Ceres.

Olbers, a doctor by profession, was an amateur astronomer, and had set up an observatory on the roof of his house in Bremen. His main work was in connection with comets, but he was also interested in the new planets, and he suggested that Ceres and Pallas might be the fragments of a larger body which had broken up for some reason. In this case there might be further pieces waiting to be discovered. In 1804 Karl Harding, Schröter's assistant at Lilienthal, found a third small planet (Juno), and in 1807 Olbers added a fourth (Vesta), which was the brightest of the four, though smaller than Ceres. At times Vesta may be seen with the naked eye as a faint speck of light.

Together, the four became known as the Minor Planets or *asteroids*. Their orbits were very similar, apart from the sharper inclination of the path of Pallas.

No more discoveries were made for some time. Schröter's observatory at Lilienthal was destroyed, and in 1815 the 'celestial police' disbanded. There the matter rested until 1830, when another amateur, Karl Hencke, returned to it. Hencke was the postmaster at his native town of Driessen, in Germany, and he felt convinced that there were more asteroids waiting to be discovered.

Alone and unaided he searched patiently for fifteen years, and at last he was successful; in 1845 he found another asteroid, now named Astræa. Less than two years later he detected No. 6, Hebe. Now other astronomers started joining in the hunt. John Russell Hind, in London, added two more asteroids in 1847 (Iris and Flora); Graham, at Markree, a ninth (Metis) in 1849; there were three discoveries in 1850, two by the Italian astronomer Annibale de Gasparis, Director of the Capodimonte Observatory near Naples, and one by Hind. Since then, not a year has passed without the addition of several more asteroids. Over 2,000 have now been studied sufficiently for their orbits to be worked out, and many more must remain undetected. One estimate gives the total number of small planets as 44,000, while Russian authorities believe that 100,000 may be nearer to the truth.

The early discoveries were of course made by ordinary visual observation—the careful checking of star-fields night after night, to see if any starlike point moved. Later, photography came into

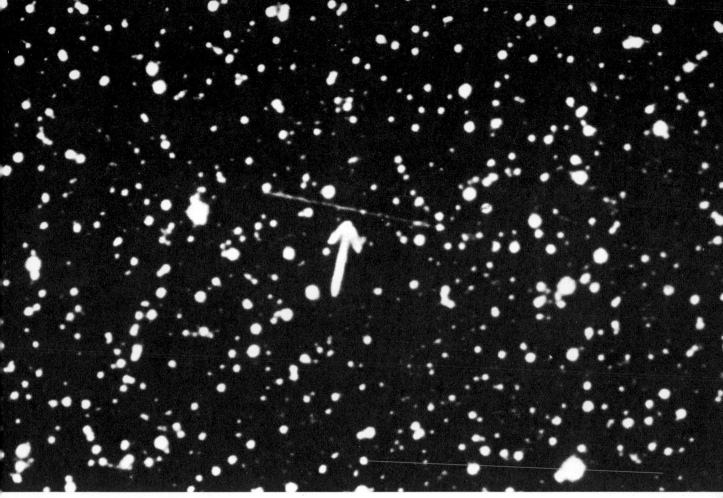

TRAIL OF ICARUS, 1949. *Icarus is an exceptionally interesting minor planet. It is not an 'Earth-grazer', but at its closest to the Sun its distance is less than that of Mercury. The movement among the stars is relatively rapid, and the trail of Icarus is indicated by the arrow; clearly the asteroid moved appreciably during the time of exposure. Photographed with the 48-inch Schmidt camera at Palomar*

use, and was responsible for revealing so many asteroids that some astronomers began to lose patience with them. One American even called them 'vermin of the skies'.

Ceres, with a diameter of 427 miles, is much the largest of the minor planets, and most of the smaller members of the family are mere lumps of material a mile or two across. None of them can have any atmosphere, and all things considered they are among the least interesting of the celestial bodies. It is still not certain whether Olbers was right or wrong in believing that they are the

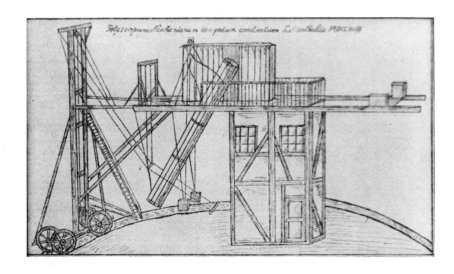

TELESCOPE USED BY SCHRÖTER. *Of Schröter's telescopes, it is possible that the largest (a 19-inch Schräder reflector) was of poor quality optically, but some of the others were undoubtedly good. One of them, made by Herschel, was extensively used for Schröter's work on the Moon and planets*

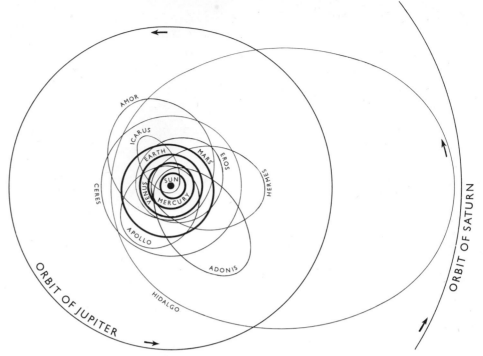

remains of a former planet which met with some disaster; it is equally likely that they represent débris left over when the main planets were formed.

Because the asteroids are so small and of so little mass, their gravitational pulls are weak. If it were possible for a man to stand upon one of the junior members of the swarm, say a mile in diameter, he would be able to jump clear of the tiny world altogether!

There are several asteroids with unusual orbits. Eros, discovered by the German astronomer Witt in 1898, comes within the path of Mars, and may approach the Earth to within 15 million miles. It is a curious body shaped rather like a sausage, 15 miles long and about 4 miles wide, and it has proved to be of real use to astronomers. In 1931 it passed by us at a distance of 17 million miles, and accurate measures of it were made, which helped considerably in measuring the scale of the Solar System and thus —indirectly—the distance between the Earth and the Sun.

Several smaller asteroids can come even closer, and in 1937 a dwarf world, Hermes, passed at a distance of only 485,000 miles, less than twice as far as the Moon. Another asteroid, Hidalgo, swings out almost as far as Saturn; Icarus moves to within 19 million miles of the Sun, so that at perihelion its surface must be red-hot. On the other hand there are two groups of 'Trojan' asteroids, named after the heroes of the Trojan War described by the poet Homer, which move in almost the same orbit as Jupiter. However, there is no fear of collision. One Trojan group keeps well ahead of the giant planet, while the other group follows far behind.

After the discovery of the first four asteroids there seemed no particular reason to suppose that any fresh planets remained to be discovered—until Uranus began to behave in a peculiar manner.

It should have been a straightforward matter to work out the way in which Uranus should move. This was particularly the case since, as we have seen, there were some early observations by Flamsteed and Lemonnier which showed Uranus clearly, even

ORBITS OF SOME INTERESTING MINOR PLANETS. *Whereas most of the minor planets revolve round the Sun at mean distances between those of Mars and Jupiter, some have more eccentric orbits which take them away from the main swarm. Icarus, for instance, has a perihelion distance less than that of Mercury; Hidalgo travels out almost as far as the orbit of Saturn. However, most of these exceptional asteroids are very small, and so are extremely difficult to observe except when fairly close to the Earth*

SIZE OF EROS COMPARED WITH MALTA. *Eros is one of the smaller asteroids, and is not even approximately spherical. Probably many of the other asteroids are of similar shape; but Eros is exceptionally easy to study, since its orbit occasionally brings it well within 20 million miles of the Earth*

though neither astronomer had recognized it as anything but a star. However, these old observations did not seem to fit. Either they were unreliable, or else Uranus must be subject to unexplained disturbances of some sort.

One of the leading French mathematicians of the early nineteenth century was Alexis Bouvard, who began his career as a shepherd boy and ended it as a famous scientist. Bouvard recalculated the orbit of Uranus, using only the observations made since Herschel's great discovery of 1781. Yet even this would not do; Uranus refused to behave, and over the years it persistently wandered away from its expected path.

Though the planets move round the Sun, each has an effect upon its companions; for instance, the disturbances or *perturbations* caused in the Earth's orbit by the gravitational pulls of Venus, Mars and other planets can be measured. Jupiter and Saturn naturally perturbed the movements of Uranus, but Bouvard had allowed for all this, and still his calculations were wrong.

In 1834 the Rev. T. J. Hussey, Rector of the Kent town of Hayes, made a most interesting suggestion. Suppose that an unknown planet were pulling on Uranus? This might account for its refusal to follow its expected path; and by working backwards, so to speak, it might be possible to track down the body responsible. Hussey went so far as to write to the Astronomer Royal, George (afterwards Sir George) Airy. However Airy's reply was not encouraging, and Hussey did no more.

In 1841 a young Cambridge student, John Couch Adams, made up his mind to attack the problem as soon as he had taken his degree. He passed his final examinations two years later, and then began to study Uranus in earnest. It was a real problem in detection. He knew the way in which Uranus was being perturbed; by sheer calculation he had to track down the planet which was causing the effects, and naturally this was far from easy.

By 1845 he had worked out where the new planet ought to be. He had no large telescope with which to search for it, and, naturally, he sent his calculations to Airy. Unfortunately Airy was still not particularly interested, and since Adams was still a young man he was reluctant to follow the matter up.

Then, in 1846, similar calculations were published by a French mathematician, Urbain Le Verrier, who had approached the problem in the same way as Adams and had reached a very similar result. A copy of Le Verrier's memoir reached Airy, who realized that it was high time to do something about the calculations which he had received many months before.

Accordingly Airy instructed two observers to start hunting in the position which Adams had indicated. One of these was James Challis, Professor of Astronomy at the University of Cambridge; the other was William Lassell, a Lancashire brewer who had become world famous as an amateur astronomer. Still there were delays. Challis had no good star maps of the area, and although he actually recorded the planet on August 4 and August 12 he failed to compare his observations. Lassell, by a stroke of ill fortune, had sprained his ankle so badly that he was unable to join in the hunt at all. Meanwhile Le Verrier's observations had been sent to Johann Encke, Director of the Berlin Observatory, and

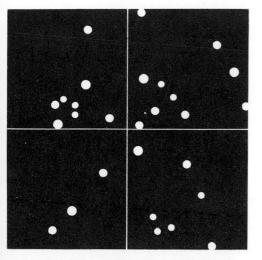

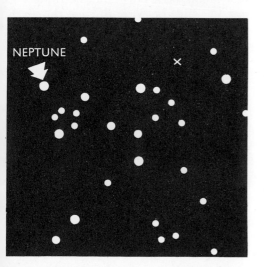

DISCOVERY OF NEPTUNE. (Upper) *Part of the star map used by Galle and D'Arrest.* (Lower) *The same map, with Neptune indicated by an arrow and Le Verrier's calculated position shown by a cross*

THE TELESCOPE WITH WHICH GALLE AND D'ARREST DISCOVERED NEPTUNE. *The telescope was a refractor made by Fraunhofer in 1820*

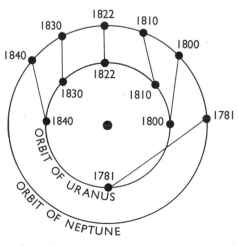

POSITIONS OF URANUS AND NEPTUNE, 1781–1840. *Before 1822, Neptune tended to pull Uranus along; after 1822, to draw it back*

Encke had at once passed them to two of his observers, Galle and d'Arrest. No time was lost. Galle searched the area with the telescope, while d'Arrest checked the stars against his chart. On September 25, 1846 they found a point of light which was not recorded on the map, and it proved to be the body for which they had been searching. It was named Neptune.

Who really discovered Neptune? Galle first identified it from Le Verrier's calculations; but Adams had finished his work earlier, and everyone now agrees that the honor should be divided equally between the Frenchman and the Englishman. It must be added, unfortunately, that the question of priority caused some bitter quarrels at the time, though neither Adams nor Le Verrier took much part in them.

It was a great achievement, and the final triumph of Newton's theories. Le Verrier had given the position of Neptune correctly to within one degree of arc, and Adams was almost as accurate.

Neptune proved to be almost a twin of Uranus—slightly larger, with a diameter of 31,200 miles, and considerably more massive. Its composition is undoubtedly similar to that of Uranus. Its average distance from the Sun is 2,793 miles, much less than expected from Bode's Law, and it goes round only once in $164\frac{3}{4}$ years. Large though it is, it is so remote that it cannot be seen without a telescope. A small instrument shows it as a tiny, rather bluish disk of about the 8th magnitude. Within a month of its discovery, Lassell—having recovered from his sprained ankle—found a satellite, Triton. Over a century later, in 1949, another much fainter moon, Nereid, was detected.

As soon as Neptune became known, the orbit of Uranus was recalculated and the extra perturbations taken into account. This time all was well, and even the old notes of Flamsteed and Lemonnier fitted into place. The diagram will help to make the position clear. Before 1822, Neptune was tending to pull Uranus along, while after 1822 it gave an opposite effect. Had Uranus and Neptune been on opposite sides of the Sun during the early nineteenth century, the perturbations would have been much less, and it is not likely that Neptune would have been tracked down.

Again the Solar System seemed to be complete, but Le Verrier at least was not so sure. He had discovered a planet beyond Uranus; was it possible that there could be another world at the opposite end of the Sun's family, within the orbit of Mercury?

Le Verrier believed so. Mercury, like Uranus, did not move exactly according to theory, and calculations showed that there might be an unknown planet pulling on it also. Le Verrier worked out its position, and even gave it a name—Vulcan, after the blacksmith of the gods.

The trouble was that Vulcan would hardly be seen under normal conditions, since it would be drowned in the Sun's glare. The only hope of finding it would be to catch it as it passed in transit across the solar disk, just as Mercury and Venus do. Consequently Le Verrier was delighted when, in 1859, he received a report from a French amateur, Lescarbault, that he had actually watched the passage of such an object. Lescarbault lived in the small town of Orgères, and Le Verrier made haste to go and see him.

NEPTUNE. *From an observation by Patrick Moore with an 18-inch reflector. Drawing by D. A. Hardy*

PERCIVAL LOWELL, *Yerkes Observatory photograph*

DISCOVERY OF PLUTO, 1930. *Pluto is indicated by arrows.* (Upper) *March 2.* (Lower) *March 5. The planet has moved appreciably in the interval. The bright, over-exposed star to the left is Delta Geminorum. Lowell Observatory photograph*

It must have been a strange meeting. By this time Le Verrier was Director of the Paris Observatory, and internationally famous, but he is also said to have been one of the rudest men who has ever lived. (Later in his career—in 1870—he was forced to resign his post, though he returned two years later when his successor, Charles Delaunay, was drowned in a boating accident off Cherbourg.) Lescarbault, on the other hand, was intensely shy, and very much of an amateur. He was the local doctor, but also a carpenter, and he used to record his observations on planks of wood, planing them off when he had no further use for them.

It is rather strange, then, that Le Verrier accepted the correctness of Lescarbault's observation almost without question. To the end of his life he continued to believe in his intra-Mercurian planet. But 'Vulcan' has never since been seen against the Sun, and it is now certain that the doctor-carpenter was wrong. The movements of Mercury have been satisfactorily explained, and Vulcan does not exist, though it is always possible that a few tiny asteroids move in orbits closer to the Sun than that of Mercury.

Finally in the story of planet-hunting we come to modern times. Even after Neptune had been found, the movements of the outermost members of the Solar System still presented problems, and by 1905 Percival Lowell, who had founded an observatory at Flagstaff in Arizona, began a search for yet another planet well beyond Neptune. Lowell is best remembered for his theories about Mars, which will be described later, but he was also a first-class mathematician, and his calculations were based on the same principles as those of Adams and Le Verrier—though in some ways they were more difficult. The expected planet failed to show itself, and was still undetected when Lowell died in 1916.

For some years the matter was dropped, but then another American, William H. Pickering, reinvestigated it and came to a similar conclusion. Again searches proved fruitless, and again the hunt was given up.

The final triumph came much later. In 1929 astronomers at Lowell's old observatory at Flagstaff started searching once more, using improved equipment, and early in the following year Clyde Tombaugh, now a famous scientist but then a young and unknown assistant, came across a very faint speck of light which proved to be the 'ninth planet'. It was named Pluto, after the God of the Underworld. Since Pluto is so far from the Sun, it must be a gloomy place and the name is an apt one.

Yet—was it such a triumph after all? When Pluto was studied, it proved to be much smaller than expected. If our measures are right, it has a diameter only about equal to that of Mars—or perhaps slightly smaller—so that it could produce no measurable perturbations in the orbits of giant planets such as Uranus and Neptune. Either Pluto is more massive than we think, or else the discovery was purely a matter of luck. So far the riddle remains unsolved.

Is there another planet beyond Pluto? Some astronomers believe that there is, and if so it may be found one day, though it is certain to be very faint. Planet-hunting today is much more difficult than in the days of the 'celestial police' and Adams and Le Verrier, but it may not be over yet.

17 great telescopes

GREGORIAN TELESCOPE, *made by James Short in 1749*

GREGORIAN TELESCOPE: *Another of Short's instruments*

UP TO THIS POINT we have been able to deal with the history of astronomy by taking each half-century or so and describing the main developments. A comparatively few famous names dominate the story: Aristarchus, Ptolemy, Copernicus, Tycho Brahe, Kepler, Newton and Herschel among others. But from the time of Herschel's death, such treatment is no longer possible, and it is necessary to take the various branches of astronomy one at a time, bringing the story up to the present day.

First, then, let us consider how telescopes and other instruments were improved.

Newton had developed the reflecting telescope mainly because he could not see how to produce a refractor which would be free of the irritating false color. The aerial telescopes of men such as Huygens and Hevelius were remarkably clumsy, and unless better lenses could be made it was quite obvious that the refractor had little future.

So matters stayed until 1729, when an English amateur named Chester More Hall, who lived near Harlow in Essex, made a series of interesting experiments and constructed the first compound or *achromatic* object-glass. Strictly speaking, such an object-glass is not one lens at all, but two—made of different kinds of glass, and placed close together. The false color produced by one lens will thus tend to be cancelled out by the other.

More Hall's theory was basically wrong. However, he did manage to build an achromatic refractor with a $2\frac{1}{2}$-inch object-glass and a focal length of only 20 inches. It was far from perfect, but it was a vast improvement on the old-type 'monsters'.

More Hall had no wish for fame, and he took no steps to make his discovery known. Not many people heard about it, and it was soon forgotten. At this period the reflectors made by men such as James Short, a famous Scottish optician, were regarded as much more promising.

Then, in 1758, an instrument-maker named John Dollond rediscovered the principle of the achromatic object-glass. Helped by his son Peter, he began to make lenses for sale, and before long refractors came back into favor. Lens-making is much more difficult than mirror-grinding, but the refractor has certain obvious advantages over the reflector, even though it is more expensive and the false color trouble can never be properly cured.

There were two great telescope-makers living in the early part of the nineteenth century: Herschel, who made considerable sums of money by selling mirrors (King George III, for instance, paid him 600 guineas each for four instruments of focal length 10 feet), and a young German, Joseph von Fraunhofer. The difference between the two was that while Herschel concentrated on reflectors, Fraunhofer was much more interested in refractors, and his ambition was to build a lens telescope which would be better than anything which Herschel could produce.

Fraunhofer was born at Straubing, in Bavaria, in 1787. Both

THE ISAAC NEWTON 98-INCH RE-
FLECTOR, *under construction at Newcastle.*
It is now in use at the Royal Greenwich
Observatory, Herstmonceux. Photograph by
Patrick Moore

PALOMAR OBSERVATORY. (left to
right) *The buildings shown are the dome of*
the 18-inch Schmidt camera; the residential
quarters; the dome of the 48-inch Schmidt;
the dome of the 200-inch Hale reflector; the
garage, and the water tower and reserve tank

his parents died while he was very young, and he had little schooling. At the age of fourteen he was apprenticed to a Munich look-
ing-glass-maker, Weichselberger. We often hear of cruel masters
and starved, ill-treated apprentices, but in Fraunhofer's case the
description fitted the facts, and the boy was desperately unhappy.
Then came an incident which altered his whole career. The
tumbledown lodgings in which he lived collapsed for no apparent
reason, and Fraunhofer was trapped in the ruins. The rescue
operations were watched by the Elector of Bavaria, who happened to be driving past, and it pleased him to befriend the lad.
He provided Fraunhofer with enough money to buy his release
from Weichselberger, and to educate himself.

Later on the Elector would have had good reason to be pleased
at the results of his good deed, since Fraunhofer became world
famous. He joined the Munich Optical Institute in 1806, and
became its Director only seventeen years later. His lenses were by
far the best produced up to that time, and when difficult problems
arose he solved them step by step. He was also the true founder of
spectroscopy, which has provided us with most of our modern
information about the stars, and which will be described in
Chapter 22. It was particularly tragic, therefore, that he died in
1826 at the early age of thirty-nine.

In 1817 Fraunhofer produced an object-glass of magnificent
quality, $9\frac{1}{2}$ inches in diameter and of 14 feet focal length. It was

DOME OF THE 102-INCH REFLECTOR AT THE CRIMEAN ASTROPHYSICAL
OBSERVATORY, *photographed by Patrick Moore in October 1960; the dome construction
was not quite complete, and scaffolding can still be seen*

72-INCH REFLECTOR AT THE KARL SCHWARZSCHILD OBSERVATORY, TAUTENBURG, EAST GERMANY. *This is an 'all purpose' reflector, using several optical systems; it can, for instance, be used as a Schmidt or a Cassegrain. It is the largest instrument of this kind yet built. Photograph by Patrick Moore, 1963*

THE 40-INCH REFLECTOR AT FLAG-STAFF. *The tube is of the skeleton type, and the observatory has a slide-back roof instead of a dome. Photograph by Patrick Moore, 1964*

bought by the Russian Government for the Observatory at Dorpat, in Estonia (which was then, as now, ruled by Russia). In 1824 the telescope was ready, and during the following years F. G. W. Struve, Director of the Observatory, discovered 2,200 new double stars with its aid. The 'great Dorpat refractor' has indeed had a distinguished history.

In another way, too, the telescope opened a new era: it was clock-driven, so that it automatically followed the stars in their east-to-west movement across the sky.

We know that the apparent daily motions of the Sun, Moon, planets and stars are not real, but are due to the Earth's rotation on its axis. Because the Earth spins from west to east, the stars seem to travel from east to west, describing paths round the celestial pole. The movement is so slow that to the naked eye it is hardly noticeable except over a period of several minutes. When one uses a telescope, however, the movement becomes very obvious. An eyepiece giving a magnification of, say, 200 will show only a very small area of the sky, and a star will seem to speed across the field, so that the telescope has to be shifted all the time to keep the object in view.

This trouble becomes worse with increasing power, and it is not easy to push a telescope smoothly enough. The remedy is to mount the telescope *equatorially*, driving it by some kind of mechanism so that it will swing round at such a rate as to keep the star firmly in the field of view.

The idea of such a driven telescope originated with Robert Hooke, but in his day equatorial mounts were not built; even Herschel's great 40-foot reflector was an *altazimuth*, so that it had to be moved up and down as well as from east to west if it were to keep a star in view. The Dorpat refractor was set up on an equatorial mount, and was driven round by clockwork. At the time, this was a daring new development. Nowadays all large telescopes are driven, though in most cases electricity has replaced clockwork.

The next great development lay not many years ahead. This was celestial photography, without which the modern astronomer would feel helpless.

The story of photography would need a complete book to itself, but it may be said to have begun with the work of men such as Fox Talbot of England, and Nicéphore de Niepce and Louis Daguerre of France, in the first half of the nineteenth century. In 1839 Jean Arago, the Director of Paris Observatory, announced the discovery of the 'Daguerreotype' process, and was quick to stress how important it would be to astronomers. He even suggested that it would become possible to obtain an accurate map of the Moon in a few minutes, instead of by years of patient observation at the eye-end of a telescope.

Early photography was very primitive, but in 1845 two French physicists, Fizeau and Foucault, took photographs of the Sun and recorded some spots. The first picture of the Moon was obtained as early as 1840, by J. W. Draper in the United States, and ten years later W. C. Bond and J. Whipple, at Harvard College Observatory, managed to photograph the two bright stars Vega and Castor.

By 1860 photography had made great strides, and was starting to be really useful in astronomical work. Today, of course, most research is carried out in this way, and the world's greatest telescopes are used mainly for taking photographs. There are several good reasons for this. Unlike the human eye, the camera cannot make a mistake, and it gives a permanent record which can be studied later on. Moreover the eye becomes tired after a few minutes' observing, while the sensitive plate goes on collecting light all the time, building up an image. It is possible to photograph objects which are so faint that they cannot be seen visually even with our largest modern telescopes.

The period of great telescopes began with Herschel's 48-inch reflector, which was not equalled in size for many years after it was first turned towards the sky. Then, in 1845, came an even larger telescope—again a reflector, this time with a 72-inch mirror made of speculum metal. It was made by an Irish nobleman, the third Earl of Rosse, and set up at his home at Birr Castle.

Lord Rosse was born in 1800. He was a rich landowner, and entered Parliament when still an undergraduate at Oxford, while from 1845 until his death twenty-two years later he sat in the House of Lords as an Irish representative peer. However, his main interests were scientific. He had to learn by experience, and it was a real triumph when he first turned his giant reflector to the skies. As the engraving shows, the instrument was clumsy and inconvenient to use—it is said that the mounting caused nearly as much trouble as grinding the mirror—but its light-gathering power was tremendous, and Lord Rosse was the first to see that some of Herschel's 'resolvable nebulæ' are spiral in form, not unlike Catherine-wheels.

The Rosse telescope has long since been dismantled. The mirror lies in the Science Museum at South Kensington; only the tube remains at the original site at Birr Castle. However, it had done its work, and it represented a new stage in the development of astronomical instruments.

The two greatest telescopes of the pre-1850 period had been built by private individuals; Herschel was an amateur until his appointment as royal astronomer following his discovery of Uranus, and Lord Rosse remained an amateur to the end of his life. But conditions were changing, and future large telescopes were confined to official observatories. When considering some of these observatories, it is natural to begin with one of the earliest and most famous of them all—Greenwich.

There have been nine Astronomers Royal since Halley died in 1742. Most of them have left their mark on the Observatory in various ways. Nevil Maskelyne, who held office between 1765 and 1811, founded the *Nautical Almanac*, which has remained the standard yearly reference book for astronomers and navigators, though it has now been combined with the leading American almanac. There was also John Pond, who followed Maskelyne, but who neglected his duties so badly that he was compelled to resign; and there was George Biddell Airy, who reorganized matters when he succeeded Pond and who obtained many new instruments.

It is often true to say that 'men's mistakes survive; their virtues do not', and Airy is probably best remembered as being the man

THE 72-INCH ROSSE REFLECTOR AT BIRR CASTLE. *An engraving of this great telescope, for many years the largest in the world*

THE WHIRLPOOL GALAXY, M.51 IN CANES VENATICI. *The spiral nature of some external galaxies was first revealed by the Rosse 72-inch reflector. Photograph taken with the 200-inch Hale reflector at Palomar*

List of Astronomers Royal

J. Flamsteed	1675–1719
E. Halley	1719–1742
J. Bradley	1742–1762
N. Bliss	1762–1765
N. Maskelyne	1765–1811
J. Pond	1811–1835
G. B. Airy	1835–1881
W. H. M. Christie	1881–1910
F. W. Dyson	1910–1933
H. Spencer Jones	1933–1954
R. van der R. Woolley	1954–1971

DOME OF THE 33-INCH REFRACTOR AT THE OBSERVATORY OF MEUDON, FRANCE. *This telescope was for many years the most powerful in Europe. Meudon lies between Paris and Versailles. Photographed by Patrick Moore, 1961*

THE LICK OBSERVATORY. *The following domes are shown* (left to right)*: the 26-inch Crossley reflector, the 22-inch Tauchmann reflector, the 36-inch reflector, the 12-inch refractor, the 20-inch Carnegie Astrograph, and the 120-inch reflector. The latter telescope is the largest reflector in the world apart from the Palomar 200-inch. Lick Observatory photograph*

DOME OF THE 50-INCH REFLECTOR AT THE CRIMEAN ASTROPHYSICAL OBSERVATORY, U.S.S.R. *Photograph by Patrick Moore, 1960*

who did *not* discover Neptune although all the information had been put into his hands. Yet this is decidedly unfair. In Airy's hands Greenwich built up its reputation, and he was a great—though severe—administrator.

Though Greenwich had been founded so that a star catalogue could be drawn up to help sailors in their longitude-finding problem, it soon became a centre of purely astronomical research. During the office of Sir William Christie, who followed Airy, a 28-inch refractor was set up, and in 1934 a 36-inch reflector—known as the Yapp, since it was presented by a wealthy enthusiast named William Yapp.

Yet one difficulty became obvious as the twentieth century passed by: Greenwich was no longer a suitable place for an observatory. In the days of Charles II, and even Queen Victoria, it had been a village outside London. Now London was spreading, and the smoke, grime and artificial lights hid the stars, so that the big telescopes could seldom or never be properly used. There was only one solution. Greenwich Observatory must move.

Near the little Sussex town of Hailsham lies Herstmonceux Castle, where the skies are clearer and there are no inconvenient factories or lights. The drastic decision was made to shift the

THE 40-INCH REFLECTOR *at the Cape Observatory, South Africa, photographed by Patrick Moore*

ARCETRI OBSERVATORY. *This photograph shows the solar dome at one of the leading Italian observatories—Arcetri, near Florence. Solar studies form a major part of the observatory programme. Photograph by Colin Ronan, 1961*

DOME OF THE 100-INCH REFLECTOR
AT MOUNT WILSON

THE 100-INCH HOOKER REFLECTOR
AT MOUNT WILSON

THE 98-INCH ISAAC NEWTON RE-
FLECTOR, *photographed by Patrick Moore
in 1965 during its assembly at Newcastle*

Observatory there, lock, stock and barrel. The move took many years to arrange, but is now complete, and though Wren's old building in Greenwich Park still stands it serves only as a museum.

If you go to Herstmonceux, you will see the old castle looking exactly as it was before the astronomers came. The interior has been completely rebuilt, and is used as offices and living quarters, while the domes housing the Yapp reflector and other instruments lie on the rising ground near by. These powerful, remarkably accurate telescopes have now been joined by the 98-inch reflector named in honor of Sir Isaac Newton. This telescope, which is by far the largest in the British Isles, was completed in 1967. The observatory in Greenwich Park had a long and honorable history; the Royal Greenwich Observatory, Herstmonceux, can look forward to a future no less profitable.

Yet the British Isles do not enjoy a good climate from the astronomer's point of view. There is a great deal of cloud, and the seeing is usually rather poor. Conditions in some parts of the U.S.A. are better, and this brings us to the great American telescopes. The genius behind the first three of them was George Ellery Hale.

Hale was born in 1868, and was fascinated by astronomy from boyhood. At the age of twenty-three he became famous for his invention of the *spectroheliograph*, a special instrument used in solar studies, and indeed he was at that time interested chiefly in the Sun.

His first observatories were privately financed by his father, a prosperous Chicago businessman, but they were too small to satisfy the young astronomer. In every way it was desirable to have a much larger telescope, and he first turned his attention to the possibilities of a big refractor. At this time the largest refractor in the world was that of the Lick Observatory, California. It had a 36-inch object-glass, so that it dwarfed the $9\frac{1}{2}$-inch Fraunhofer refractor which had been erected at Dorpat so long before.

Hale wanted a 40-inch refractor, but naturally he could not afford anything of the sort. Only a few men would have been capable of making such a lens. The greatest expert in the world was Alvan G. Clark, also an American, but even to him the task would take years—and there was also the bill for the glass blanks, which alone would cost 20,000 dollars.

In 1892 Hale met Charles T. Yerkes, who was immensely rich and who owned a large part of the city of Chicago. Yerkes could afford to pay for the telescope—and he did; the total cost came to 34,900 dollars. Clark ground the mirror, and the Yerkes refractor was duly completed. In 1897, the observatory was opened. It lay at Williams Bay, on the shores of the lake about 70 miles from Chicago, and contained many instruments as well as the main telescope. Hale of course became Director, and within a few years the scientific results had more than justified the cost.

There is a limit to the size of a useful lens. Anything larger than a 40-inch will begin to bend under its own weight, since it has to be supported round its edge (in fact a 48-inch object-glass was once made in France, and mounted, but proved to be of little use). The Yerkes refractor remains, and possibly always will remain, the

THE 200-INCH HALE REFLECTOR AT
PALOMAR. *This, the largest telescope in the
world, has some special features possible only
because of its great size. The observing cage
lies at the prime focus, and in the photograph
an observer is shown—so that the astronomer
is situated inside the telescope. The reflecting
surface of the main 200-inch mirror is also
shown. The same arrangement will be
adopted with the 236-inch reflector now under
construction in the U.S.S.R.*

DOME OF THE 98-INCH REFLECTOR
*at Herstmonceux, photographed by Patrick
Moore*

largest lens telescope in the world. To collect even more light, it is
necessary to go back to the reflector. A mirror, remember, has
only one reflecting surface, and can be supported at the back.
Moreover it is easier to grind than a lens of equal light-grasp.

Hale's constant wish was: 'More light!' To explore the depths
of the universe—the remote star-systems—was beyond even the
Yerkes refractor, and accordingly Hale searched round for the
money to build something even better. Again he was lucky. A
great financial trust known as the Carnegie Foundation, so called
after Andrew Carnegie, one of the few men as wealthy as Charles
Yerkes, promised to provide a 60-inch reflector. George Willis
Ritchey, unrivalled as a telescope-maker at that period, took
charge of the mirror-grinding.

One important problem concerned the choice of a site. The
Earth's atmosphere becomes thinner with increasing altitude, and
since the dirtiness and unsteadiness of the air is probably the
astronomer's worst enemy it is common sense to erect a large
telescope on the top of a mountain. Two peaks in California
seemed to suit all the requirements—Mount Wilson, and Palomar.
Palomar was probably the better, since it was farther from any
artificial city lights, but on the other hand it was very hard to get at,
and so Hale and his colleagues decided upon Mount Wilson. The

THE 26-INCH REFRACTOR AT THE U.S NAVAL OBSERVATORY, WASHING-TON. *This telescope is one of the great refractors erected during the latter part of the nine-teenth century. At this period large refractors were generally preferred to large reflectors; lens-making had reached a high degree of perfection, whereas the mirror-making methods in use today had not been developed.*

The 26-inch has been used for a variety of purposes, both research and educational. For instance, extensive series of double star measures have been made with it. Its most famous association is with the two dwarf satellites of Mars, Phobos and Deimos, since this was the telescope with which Asaph Hall discovered them in 1877

MOUNT STROMLO OBSERVATORY, AUSTRALIA. *At present the largest reflector in the southern hemisphere is at Mount Stromlo, at Canberra in Australia (the Commonwealth Observatory). The mirror is 74 inches in diameter; the dome is shown in this photograph. The moving parts of the telescope weigh 40 tons, while the revolving dome has a weight of 100 tons—and yet can be moved smoothly on its track at the rate of one complete revolution in five minutes. Because of the southern latitude, astronomers at the Commonwealth Observatory can study objects such as the Clouds of Magellan which can never be seen by their colleagues at Palomar or Mount Wilson*

24-INCH REFLECTOR *at the Kwasan Observatory, Kyoto, Japan, photographed by Patrick Moore*

60-inch telescope came into use in 1908, and more than fulfilled the hopes of its builders.

Yet even before then, a larger telescope had been planned. The suggestion came from another millionaire, John D. Hooker of Los Angeles, who went to see Hale and asked whether it would be possible to build a 100-inch reflector. Hale was sure that the answer was 'yes', and Hooker promised to provide the vast sum of 45,000 dollars—enough to pay for the mirror, though not for the mounting, which also was bound to be very expensive.

Money was only part of the programme; the work remained to be done. Casting the glass disk caused many anxieties even before the actual grinding could be begun, and to shape the mirror to the correct optical curve took Ritchey and his team six years. Meanwhile Andrew Carnegie had visited Mount Wilson, and had been so impressed that he had agreed to meet the cost of mounting the 100-inch mirror.

One evening in November 1917, while the First World War was raging in Europe, Hale and three companions turned the Hooker reflector to the skies for the first time. They focused carefully on the planet Jupiter, and were appalled at what they saw—a shimmering, blurred image, lacking in any sort of detail. Could the mirror be useless after all, in spite of all Hooker, Carnegie and Ritchey between them had done?

Hale was not sure. The dome had been opened during the afternoon, and the mirror had become warm. Large glass disks take a long time to cool down and regain their correct shape—and a distortion of a tiny fraction of a millimetre would be too much. Hours later Hale went back to the telescope, and turned it towards the star Vega. This time all was well, and he knew at last that the telescope was as perfect as even he and Ritchey could have wished.

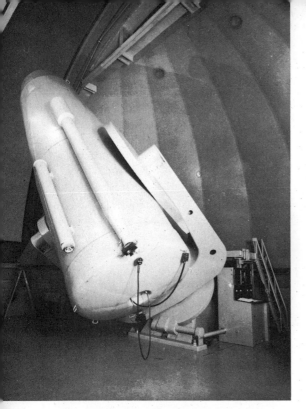

THE 48-INCH SCHMIDT AT PALOMAR.
A Schmidt telescope, perhaps more appropriately termed a Schmidt camera, can photograph relatively large areas of the sky with excellent definition. Instruments of this sort cannot be used for visual observations

PATRICK MOORE'S 5-INCH REFRACTOR. *An amateur-owned telescope, at Selsey, photographed by Patrick Moore*

It would be difficult to over-estimate how valuable the Hooker reflector has been to astronomy. It opened up entirely new fields of research; it was so much more powerful than any other instrument that it was in a class of its own. With it, for instance, Edwin Hubble proved that Herschel had been right in suggesting that the 'starry nebulæ' were separate systems far beyond our Galaxy. It more than doubled mankind's knowledge of the universe.

Last and so far the greatest of these huge reflectors is the 200-inch at Palomar Mountain, also in California. Again the idea was Hale's, though unhappily he died in 1938 ten years before the instrument was finished. Again there were immense difficulties to be faced, both financial and practical; if making the 100-inch mirror had been a hard task, the problems of the 200-inch were even greater. All were surmounted, and it was only fitting that the telescope should be named the Hale Reflector. At the opening ceremony on June 3, 1948, Dr. Lee Du Bridge, President of the Carnegie Institute, spoke some words which deserve to be remembered: 'This great telescope before us today marks the culmination of over 200 years of astronomical research. For generations to come, it will be a key instrument in man's search for knowledge.'

Galileo and Newton, even Herschel and Fraunhofer, would have been staggered by the Hale reflector; it is so large that the observer sits in a cage slung inside the tube, and all movements of the telescope are controlled by complex electric mechanisms. It, too, has provided an immense amount of information which could never have been gained without its help.

Yet how many people have much idea of what a modern observatory is like? The general impression is of an isolated dome on the top of a mountain, where an astronomer sits night after night gazing through a telescope and making notes, getting his sleep during the daytime (unless, of course, he is studying the Sun).

Nothing could be further from the truth. An observatory such as Palomar is almost a city in itself. There are machine shops, living quarters, libraries, lecture halls, chemical and physical laboratories, and photographic stores and dark-rooms; there are instruments of all kinds as well as the main telescopes. A telescope such as the Hale reflector is almost never used visually. Its main role is to take photographs, and the chief attention is riveted upon the distant star-systems.

Neither does an astronomer continue photographing for night after night. He may spend a week at the telescope, after which he will go back to 'civilization'—the town of Pasadena, if he has been working at Palomar—and spend weeks or even months studying his pictures and analyzing the results, while another observer takes his place in the observatory itself. Even when the photographs have been taken, the main work has only just begun.

Other huge telescopes have come into use. For·instance, the 36-inch refractor at Lick Observatory, on Mount Hamilton, has been joined by the 120-inch reflector completed in 1958, while the McDonald Observatory, in Texas, has an 82-inch reflector dating from 1939.

Europe and the United States lie to the north of the equator, and some of the southern stars never rise. This is unfortunate, since

DOME OF THE 50-INCH REFLECTOR
AT THE CRIMEAN ASTROPHYSICAL
OBSERVATORY, *taken from the dome of
the adjacent 102-inch by Patrick Moore,
1960*

DOME OF THE 74-INCH REFLECTOR
AT THE COMMONWEALTH OBSERVA-
TORY, *at Mount Stromlo in Australia—the
largest reflector in the southern hemisphere*

these regions contain objects of tremendous interest. It would be valuable to have a 200-inch reflector in, say, Australia or South Africa; meanwhile there are major projects under way in both these countries, and in South America. It is no longer necessary to take telescopes from the northern to the southern hemisphere, as Edmond Halley and Sir John Herschel had to do.

Lastly, what of Russia?

Things have changed since the time of Lomonosov, and to-day astronomers in the Soviet Union are among the most advanced in the world. Their equipment, too, is as good as any. It was a disaster when one of their best observatories, Pulkovo, was shelled by the Germans during their attacks on Leningrad during 1941; every dome was destroyed, and the observatory had to be completely rebuilt, though it is now in full working order again. Among other Russian observatories are those at Taskhent, Abastumani and Burakan.

The Crimean Astrophysical Observatory is favoured by good seeing conditions, and a 102-inch reflector has been set up there beside the older 50-inch reflector. Moreover the Soviet authorities are now testing a 236-inch telescope, which is larger even than the Hale.

Will this Russian reflector remain the 'biggest ever'? It is hard to say. The Earth's atmosphere is a crippling handicap; the larger the telescope, the worse the troubles become, and it may be that if we are to build yet more powerful instruments we will have to set them up either out in space, or on the surface of the Moon. This would have seemed fantastic even a decade ago, but it may well have been achieved before the end of the twentieth century.

This description of modern observatories is very incomplete, but it may give an idea of how an astronomer of today carries out his work. He can make use of all the resources of science; he can call upon engineers, chemists, physicists and many others, and he is in constant touch with his colleagues all over the world. If a discovery is made at Palomar, then astronomers in Herstmonceux, Leningrad, Canberra or Tokyo will hear all about it within an hour or two.

We have come a long way since the days of Galileo.

18 explorers of the Moon

THE LUNAR CRATER TYCHO. *The central peak is a prominent feature. Lick Observatory photograph*

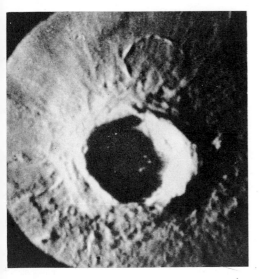

THE LUNAR CRATER ARISTILLUS. *A prominent crater, photographed at the Pic du Midi Observatory*

◄ THE VOLCANIC CRATER OF HVERF-JALL, IN ICELAND (*upper*). *This vulcanoid, close to Lake Mývatn, bears a strong superficial resemblance to a lunar crater; the central peak is a prominent feature. Photograph by Brian Gulley, 1960.* THE ARIZONA METEOR CRATER (*lower*), *photographed from the air by Patrick Moore. This is undoubtedly due to the impact of a large body which hit the desert in prehistoric times, and is almost one mile in diameter. No doubt many of the smaller craters on the Moon were produced in this way, but whether the larger lunar craters are volcanic or meteoric is still a matter for debate.*

GALILEO COMPLETED his first telescope towards the end of 1609. It was natural for him to turn it to the Moon, since even with the naked eye it is possible to see light and dark patches on the lunar disk. Moreover the Moon is much the closest of all celestial bodies, and is the Earth's companion in space.

By Herschel's time a great deal of information had been collected. It was known that the Moon is relatively small, with a diameter of 2,160 miles, and has only $\frac{1}{81}$ of the mass of the Earth; that its surface contains mountains, valleys and craters; that part of its face is permanently turned away from us, and that the so-called 'seas' are not seas at all. It was also believed that the Moon has almost no atmosphere.

Herschel himself firmly believed the Moon to be inhabited, which meant that 'air' could not be completely lacking. However, Herschel's main attention was concentrated on the stars, and the true founder of *selenography*, the physical study of the Moon, was Johann Hieronymus Schröter.

Schröter was seven years younger than Herschel. He was born at Erfurt, in Germany, and studied law at the University of Göttingen. He became a civil servant, and was appointed chief magistrate of the town of Lilienthal, near Bremen. Here he set up an observatory, and obtained one of the telescopes made by William Herschel. Then, in 1792, he was visited by J. G. F. Schräder, Professor of Mathematics and Physics at the University of Kiel. Schräder was handicapped by his serious deafness, but he and Schröter had much in common; moreover Schräder was a telescope-maker, and he produced a 19-inch reflector which remained in use at Lilienthal for many years.

Lilienthal became a recognized astronomical centre. It was here that the 'celestial police' met to organize their hunt for the supposed planet between Mars and Jupiter, and later on Schröter had various assistants, among them Karl Harding—discoverer of Juno—and Friedrich Bessel, who achieved lasting fame in later years as being the first man to measure the distance of a star.

Schröter's work did not overlap that of Herschel. He was concerned almost entirely with the Solar System, and particularly with the Moon. He never produced a complete lunar map, and the chart by Tobias Mayer, which had appeared in 1776, remained the best until 1838. Yet Schröter made hundreds of careful drawings, and discovered many new lunar features. He was the first to give a proper account of the curious *clefts* or *rills*, which look like cracks in the surface; indeed, the only previous record of them was contained in one observation made by Christiaan Huygens. Later selenographers were able to build upon the foundations which Schröter had provided.

It is rather strange to find that the value of Schröter's work has often been questioned. It is said that his drawings were clumsy, his telescopes poor, and his theories absurd. None of these criticisms is fair. He was not a good draughtsman, but he seldom made a

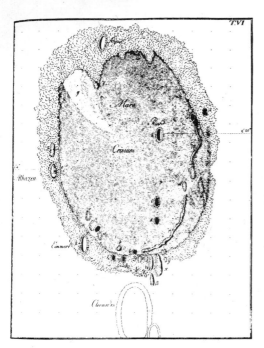

THE MARE CRISIUM, AS DRAWN BY SCHRÖTER. *This is a typical drawing by Schröter, who was the first of the great lunar observers. It is said, and with justification, that Schröter was a clumsy draughtsman; nevertheless he seldom made a serious mistake, and the details are always clearly shown, as may be seen by comparing this drawing with a modern photograph. It shows the Mare Crisium or Sea of Crises, near the western limb of the Moon—one of the smaller but most conspicuous of the dark plains*

serious mistake—and though the 19-inch reflector may not have been first-class, the smaller telescope made by Herschel was certainly excellent. Schröter was wrong in thinking that the Moon had an appreciable atmosphere, and that changes occurred on the surface, but his ideas were no more extreme than Herschel's.

Schröter's work came to an unhappy end. During the Napoleonic Wars, in 1813, his observatory was burned down, together with all his unpublished observations. Even his telescopes were plundered, and all that Schröter could do was to save as much as he could. He began work on a new observatory in Göttingen itself, but he was now an old man, and the effort was too great. His life was wrecked, and in 1816 he died. The remains of his observatory at Lilienthal were dismantled in 1840, and nothing now remains there apart from a memorial tablet.

All those who examine Schröter's books will be left with a feeling of deep admiration for him. He was a real pioneer, and one of the greatest of amateur astronomers.

Meanwhile, a really good map of the Moon was badly needed. Wilhelm Lohrmann, born at Dresden in 1796 and trained as a land surveyor, determined to draw one; he completed four sections of it, but unfortunately his eyesight failed, and he had to give up.

At about the same time, Johann von Mädler, a Berlin teacher, was giving private lessons in astronomy to a wealthy banker named Wilhelm Beer. Beer was an apt pupil, and before long the two men joined forces. Beer equipped his observatory with a fine $3\frac{3}{4}$-inch refractor made by Fraunhofer, and together they began work on a lunar chart. It appeared in 1837–8, together with a book, *Der Mond* ('The Moon'), containing a description of all the important mountains and craters.

The map was remarkably good, and a triumph of patient, skilful observation. It was regarded as 'the last word' on the subject, but the results of its publication were unexpected. For some time afterwards most astronomers neglected the Moon altogether, and turned their attention elsewhere.

The reason for this was simple enough. Beer and Mädler

THE HYGINUS CLEFT, *photographed at the Pic du Midi Observatory. This is a prominent feature, visible in a very small telescope. Under low magnification it appears as a true crack or cleft, but close examination with a higher power shows that it is, in part at least, a crater-chain and not a genuine cleft at all. In this photograph, many of the craterlike enlargements are shown. Hyginus itself lies in the middle of the picture*

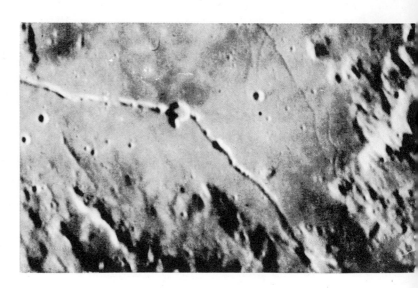

TYCHO. *Tangled rocks outside the wall of the crater, photographed from the Surveyor craft which landed there*

ALPETRAGIUS. *One of the great lunar craters, photographed from Orbiter*

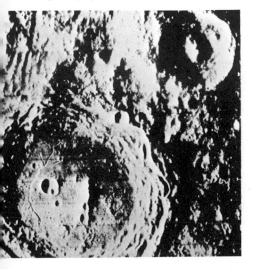

believed the Moon to be a dead world—barren, lifeless, almost without atmosphere, and completely changeless. In their view, nothing had happened on the Moon for millions of years, or would do so for millions of years to come. They had charted the surface as well as possible with their Fraunhofer refractor, and other observers felt that there could be nothing more left to be done.

This sort of attitude sounds strange today, but it appeared quite rational in 1838! Shortly afterwards Mädler left Berlin, and never undertook much more lunar work; Beer, too, was content with casual observing. Their map remained unrivalled for many years, and in the mid-nineteenth century the only really serious lunar worker was Julius Schmidt, a German who spent much of his life in Greece and become Director of the Athens Observatory. Schmidt became interested in the Moon while he was still very young, and in 1866 made the spectacular announcement that a small crater named Linné, on the well-marked Mare Serenitatis (Sea of Serenity) had disappeared, to be replaced by a craterlet in the center of a white patch.

Tremendous interest was generated, and astronomers began to turn their telescopes back toward the Moon. Even today there is some argument as to whether or not a true change had taken place in Linné; on the whole the evidence seems against it, but the episode was highly beneficial to selenography. In England, Edmund Neison published a good lunar map in 1876, based on Beer and Mädler's; two years later Schmidt produced an excellent and elaborate chart, which was really a completion and improvement of Lohrmann's. From 1895, the Lunar Section of the newly-formed British Astronomical Association took up the work, and under successive highly-skilled Directors—T. G. Elger, then Walter Goodacre, and then H. P. Wilkins—was responsible for a vast amount of detailed lunar work. Wilkins, who was born in 1896, went so far as to compile a map 300 inches in diameter. He lived just long enough to see the first rockets reach the Moon; he died in January 1960.

Meantime, lunar photography had been making progress. The first pictures were taken in 1840, and the first good photographic atlases appeared before the end of the century; that constructed by Loewy and Puiseux, at Paris, was the best of them. However, the Moon was still regarded as mainly an amateur province, and photographs of it taken with large telescopes were in general both casual and unsystematic. It was only after the end of the Second World War, when space research ceased to be regarded as the dream of a few idealists, that the whole situation changed. Professional astronomers entered the field; really good lunar photographs were taken, particularly at the Pic du Midi Observatory, 10,000 feet high in the French Pyrenees, and by an American team led by G. P. Kuiper. The Moon was becoming of more than academic importance. And then, in 1959, came the first successful lunar probes.

Because the story of selenography is so sharply divided into two periods—pre-1959 and post-1959—it may be helpful to pause here in order to give a brief summary of what had been learned from sheer telescopic observation. After all, this has provided a basis for everything which has followed, and some of the earlier

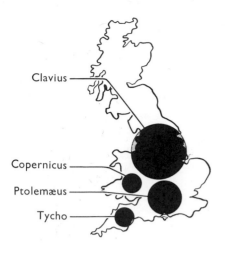

Clavius

Copernicus

Ptolemæus

Tycho

SIZES OF SOME LUNAR CRATERS,
compared with the British Isles

THE PICTURE OF THE CENTURY
*(lower). The lunar crater, Copernicus, photo-
graphed from Orbiter 2.* THE MOON FROM
ORBITER 4 *(upper). Much of the area
shown is permanently turned away from
Earth*

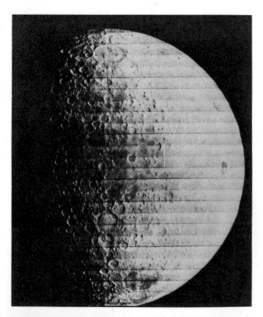

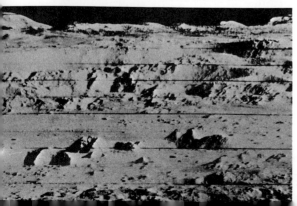

theories have been found to be very close to the mark.

Any casual glance will show that the surface of the Moon is totally unlike that of the Earth, and the reason is not far to seek: the Moon has no atmosphere. This means that it must also lack water, and the huge gray areas which are still known as seas, or maria, are not watery; there is no trace of moisture anywhere on the Moon. The romantic names, such as Mare Imbrium (Sea of Showers), Mare Tranquillitatis (Sea of Tranquillity), Sinus Iridum (Bay of Rainbows) and Oceanus Procellarum (Ocean of Storms) are never likely to be altered; most of the main features were named by Riccioli in 1651, and the system has since been extended to smaller features as well as to objects on the Moon's far side. Yet for well over a century the harsh nature of the lunar environment has been known.

The maria are much smoother than the bright uplands, and many of them are more or less circular in form, though the effects of fore-shortening make them seem elliptical when viewed from Earth. The greatest of them, the Oceanus Procellarum, is considerably larger than our Mediterranean. They are by no means featureless, and the more regular maria are mountain-bordered. Thus the lunar Alps and Apennines form part of the boundary of the well-marked, regular Mare Imbrium. Terrestrial-type mountain ranges are absent, but there are vast numbers of isolated peaks and clumps of elevations. The highest mountains on the Moon rival our Everest; their altitudes have been measured by the shadows which they cast on to the ground below, though the values given have to be reached from a mean surface level (there is no sea-level!) and are bound to be somewhat arbitrary.

The craters dominate the entire lunar scene. They range from huge enclosures well over 100 miles in diameter down to tiny pits, and no part of the Moon is free from them. They cluster in the highlands, and are also to be found on mare floors and even on the slopes and crests of peaks. They break into each other and deform each other; also they are obviously of different ages, and some of them are so ruined as to be almost obliterated. Such is Stadius, near the magnificent 56-mile crater Copernicus. Stadius has walls which can be little more than 30 feet in height anywhere, and it gives the impression of having been overwhelmed by lava rolling from the mare surface, though different astronomers have different ways of interpreting things.

The craters are of various types. Some have high, terraced walls and lofty central peaks or groups of peaks; some are low-walled, with floors which are comparatively smooth. Aristarchus, 23 miles across, is so reflective that it can be seen even when lit only by earthshine; the 60-mile Plato has a floor so dark that Hevelius, in 1645, named it 'the Greater Black Lake'. However, well-developed craters have several features in common. They are essentially regular in form, and their floors are sunken with respect to the outer surface; if there is a central elevation, the height never attains that of the encircling rampart. Moreover, the wall slopes are surprisingly gentle, and if drawn in profile a crater is seen to be much more like a shallow saucer than a deep, steep-sided mine-shaft. An astronaut who lands inside a major crater need have no fear of being shut in by towering walls. In many cases the walls

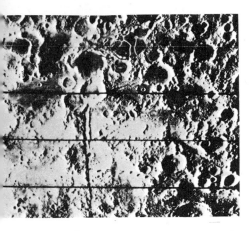

THE SCHRÖDINGER VALLEY, *a great valley on the Moon's far side, photographed from Orbiter*

THE FAR SIDE OF THE MOON, *from Orbiter 2, showing the southern area. The far side contains no major 'seas'*

would not even be visible; they would be below the observer's horizon. Remember, the Moon is a smaller world than the Earth, and its surface curves much more sharply, so that the horizon is closer.

A lunar crater appears at its most spectacular when seen near the terminator, or boundary between the sunlit and night hemispheres. It is then shadow-filled and seems most imposing, whereas when the Sun has risen high over it, and the shadows have shortened, it may be hard to identify at all. This is one reason why the best views of the Moon, telescopically, are obtained during the crescent, half and gibbous stages. At or near full, the shadows almost disappear, and the overall effect is one of confusion— particularly as some of the craters, such as Tycho in the southern part of the Moon and Copernicus and Kepler in lower latitudes, are the focal points of systems of bright streaks or rays which overlie the other features and tend to mask them. The rays are surface features; they cast no shadows, and are not well seen except under high illumination.

Of the less obtrusive features, the clefts (alternatively known as rilles or rills) are of special interest. They look like cracks in the surface, and are relatively deep. Some of them, such as the majestic winding valley near Aristarchus and the 150-mile cleft close to the little crater Ariadæus, can be seen with small telescopes; some, such as the Hyginus Cleft, are basically made up of chains of small craters which have 'run together'. There are areas which are criss-crossed with clefts, such as the interior of the imposing crater Gassendi at the border of the Mare Humorum (Sea of Humours). Then there are the low swellings known as domes, many of which are crowned by summit craterlets; there are faults, rifts and many other features. Note also the Straight Wall, in the Mare Nubium (Sea of Clouds) which is mis-named, since it is not straight and is not a wall. The surface drops sharply by 800 feet, making an incline with an angle of about 40 degrees. Before full moon it casts a shadow, and appears as a black line; after full moon the sunward face is illuminated, and the 'Wall' appears bright.

For well over a century now there have been violent arguments as to the origin of the craters. Discounting some strange theories (coral atolls, for instance, and the results of nuclear war!) there are only two really plausible ideas. Either the craters are due to internal action, so that broadly speaking they are volcanic, or else they are due to the impacts of meteorites. Both ideas have their supporters, and even today the problem has not been cleared up. It seems definite that there must be craters of both types; but which force played the dominant rôle? Largely because of the non-random distribution of the craters (together with the circular maria, which seem to be due to the same process) my view is, and always has been, that internal forces rather than meteoritic bombardment have been chiefly responsible for molding the Moon's surface, but the last word has by no means been said.

Another controversy concerned the possibility of minor activity on the Moon. Before 1958 the observers were strongly at variance with the professional theorists. Officially the Moon was regarded as totally inert and changeless; many observers did

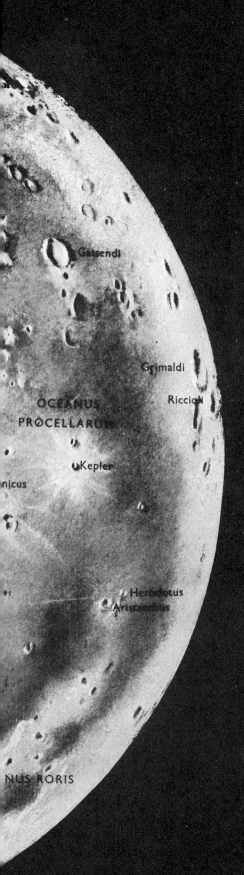

Gassendi

Grimaldi

OCEANUS
PROCELLARUM

Riccioli

Kepler

nicus

Herodotus
Aristarchus

NUS RORIS

Map of the Moon

This chart has been based on two half-moon photographs. The shadows are therefore not consistent, but the method enables most of the important craters to be shown in recognizable form. The features near the limb are not so clearly marked, since both the photographs concerned were taken under fairly high light for these areas, but various special features, such as the dark-floored craters Grimaldi and Riccioli, can be made out. In the map, the most prominent features are named, with the craters lettered in script and other objects (such as mountains) in capitals. For the 'seas', the Latin names are used, so that for instance the Sea of Showers becomes the 'Mare Imbrium'. This is by far the best method, since the Latin names are international

TYCHO, *photographed from Orbiter. The crater is 54 miles in diameter*

THE SCHRÖTER VALLEY *(upper). High-resolution Orbiter 5 picture of part of the valley.* TSIOLKOVSKII *(lower), the great crater on the Moon's far side, photographed from Orbiter. Note the blackness of the floor, which is due not to shadow, but probably to lava*

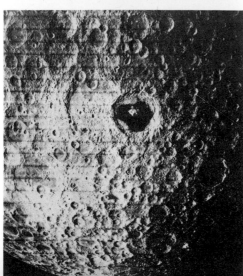

not agree, simply because they had seen the phenomena which w now call T.L.P.s (a convenient abbreviation for Transient Luna Phenomena; so far as I know I was the first to use the term, but i has now become common). The phenomena were elusive an short-lived, usually in the form of reddish patches over limite areas, and attributed to gaseous emissions from beneath th Moon's crust. Certain areas, such as Aristarchus and Gassendi were particularly subject to them.

Soon after the war, I made the prediction that as soon a professional observers started looking for these transient glows they would duly see them. I was right. The first case was that o November 3–4 1958, when a reddish event in the large crate Alphonsus was seen by the Russian astronomer N. A. Kozyrev using the 50-inch telescope at the Crimean Astrophysical Observa tory. Since then there have been many more cases, and it ma justly be claimed that the pre-Space Age observers—many o them amateur—have been fully vindicated. The emissions ar very minor by terrestrial standards, but they do exist. Moreove investigations by Jack Green and Barbara Middlehurst in th United States and myself in Britain have established that th glows are commonest near the time of perigee—that is to say, th time when the Moon is at its closest to the Earth, and the crust i subjected to the greatest gravitational strain. On the other hand major structural alterations in the surface features do not occu There has been much talk about an alleged change in Linné, o the Mare Serenitatis, which was described as a deep crater b early observers (before 1866) and is now a small craterlet sur rounded by a white nimbus; but the evidence is very slende indeed. The Moon's really active period ended long ago, probabl in the times which on the terrestrial geological scale we shoul class as Pre-Cambrian.

By 1959 there were no doubts about the unfriendliness of th Moon. As measurements became more and more accurate, th lack of atmosphere became more and more apparent. There is a easy way to show that the Moon cannot have dense 'air'; when passes in front of a star, and hides or occults it, the star snaps out a quickly as a candle-flame in the wind, whereas if the lunar lim were surrounded by atmosphere the star would flicker and fad before vanishing. In 1949 Y. N. Lipski, in Russia, announced tha by an indirect method he had found a lunar atmosphere with ground density of about 1/10,000 of that of the Earth; but it now certain that the results were erroneous. The Moon has n permanent atmosphere worth mentioning. This means also tha the temperatures are extreme—as was first established in th nineteenth century by the fourth Earl of Rosse. At maximum hea on the equator, a thermometer would register over +200 degree Fahrenheit; at night, the temperature must fall to over 25 degrees Fahrenheit below zero.

One fact which was not known was the firmness, or frailty, the surface. In 1954 Thomas Gold proposed that the maria a least must be covered with deep, soft dust, so that any space craft incautious enough to land there would sink out of sight wit devastating permanence. To observers such as myself this theor seemed to fly in the face of the evidence, as we said forcefully at th

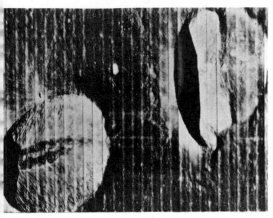

MESSIER AND MESSIER A. *The twin craters photographed from Orbiter 5, 1967. Each crater is less than 10 miles in diameter*

THE SCHRÖTER VALLEY *(upper). Part of the great feature extending from the crater Herodotus (near Aristarchus), photographed from Orbiter 5, 1967.* LUNA 13 LUNAR PANORAMA *(lower). Part of a picture transmitted by Luna 13 in December 1966*

time; but the idea was taken very seriously by the space-planners, and clearly it would be necessary to make a good many automatic landings before risking the life of an astronaut in an attempted Moon voyage.

This, then, was the state of affairs when the Russian scientists involved in rocket research opened their assault on the Moon.

In January 1959 they achieved what the Americans had failed to do in the previous year: they sent a probe, Lunik 1, past the Moon. The minimum distance was 4,660 miles, on January 4. There was no landing, but interesting data were sent back—notably to the effect that the Moon's magnetic field was extremely weak; Lunik 1 did not detect it at all. On September 12 of the same year Lunik 2 crash-landed in the Mare Imbrium. And on October 4, exactly two years after the launching of the first artificial satellite, Lunik 3 began an epic journey which took it round the Moon.

Lunik 3 was designed as a photographic probe. On October 7 it had passed by the Moon and lay 35,000 miles beyond, facing the side which is always turned away from the Earth. Photographs were taken; and when the Lunik had swung back toward Earth, later in the month, the results were sent back to the ground operators. Only part of the far hemisphere was covered, and by modern standards the pictures were blurred in the extreme; but they did show that major maria were absent, and that there were numbers of craters. In every way Lunik 3 was a triumph, and it ushered in a new era of lunar exploration. Names were given to the features whose existence was established; we still know the dark plain of the Mare Moscoviense (Moscow Sea) and the majestic, dark-floored crater which was appropriately christened in honor of the 'father of space-travel', Konstantin Tsiolkovskii.

There followed a lull, during which time the Americans managed to solve many of their worst rocket problems. The U.S. Ranger programme was designed to crash probes onto the Moon, using the last few minutes of flight to obtain and transmit close-range pictures of the surface. There were six failures; but then, on July 31, 1964, came success with Ranger 7, which came down in the Mare Nubium and sent home over 4,000 photographs before smashing itself on the bleak lunar rocks. The pictures from it, and from its successors—Ranger 8 (February 1965) and Ranger 9 (March 1965)—were far better than those from Lunik 3. They showed details which were much too small to be visible from Earth even with our most powerful telescopes, and they carried out their missions without a hitch.

Then, on January 31, 1966, the Russians launched Luna 9. (The old term 'Lunik' had been tacitly replaced by 'Luna' after No. 3. Lunas 4 to 8 inclusive failed for various different reasons.) Luna 9 came down in the Oceanus Procellarum, but it did not crash; it made a successful soft descent, so that after arrival it was able to go on transmitting photographs as well as general data. Oddly enough, the first pictures to be released were those received at Jodrell Bank, in England, with improvised equipment borrowed from the *Daily Express* newspaper! The Soviet photographs followed soon afterwards, and showed a scene which was fairly obviously a lava-plain. Gold's deep-dust theory was finally dis-

FOOTPRINT ON THE MOON, *Apollo 11, July, 1969. The print of the astronaut's boot is very plain*

EAGLE, FROM COLUMBIA *(upper). The Lunar Module of Apollo 11 above the Moon, photographed by Michael Collins from the Command Module.* MOON-ROCK *(lower). One of the first pieces brought back by Apollo 11. This was photographed by Patrick Moore at Houston, Texas, shortly after the return of the space-craft. The rock is enclosed in a glass dish, as it was still then in 'quarantine'*

proved. Luna 9 showed no inclination to sink out of view; it had come to its final resting-place upon a surface which was pleasingly hard. Two months later, the Russians achieved another 'first' when Luna 10 was put into a circum-lunar path, thereby becoming the first satellite of a satellite.

The American Surveyor probes followed. There were seven of them altogether, of which the odd numbers (1, 3, 5 and 7) were successful. Like Luna 9—and also Luna 13, of December 1966—they made soft landings, and sent back thousands of fine-quality photographs. The last of the series, Surveyor 7 of January 1968, landed in the high ground just outside the ray-crater Tycho, and sent back pictures of a scene which was wild in its grandeur. Yet of even more importance were the five Orbiters, beginning in August 1966 and ending in August 1967. All functioned perfectly; all went into closed paths round the Moon, and transmitted photographs which enabled the American astronomers to put together an amazingly detailed map of almost the entire lunar surface. Over a century and a quarter earlier, in 1840, the French astronomer Jean Arago had said that photography would result in the compilation of a really accurate, large-scale chart of the Moon. He was right; but the map was completed only with the work of Orbiter 5.

Obviously, all this research was leading up to the supreme experiment: that of landing men on the Moon. Around 1960 it was generally assumed that the first lunar pioneer would be a Russian, but as time went by it became clear that the Soviet plans were being concentrated upon automatic probes. Meantime, Project Apollo went ahead. This is not the place to enter into a description of the techniques involved (see my *Space*, Lutterworth Press, fourth edition 1971), and in any case most people must surely have seen at least some of the Apollo manœuvres on television. Suffice to say that after the preliminary tests had been made, the astronauts of Apollo 8—Frank Borman, James Lovell and William Anders—made the first circum-lunar journey. They went round the Moon ten times at Christmas 1968, and so became the first men to have a direct view of the Moon from close range. Their minimum distance from the surface was a mere 70 miles. Apollo 9 was an Earth-orbiter; Apollo 10, launched on May 18, 1969, carried Astronauts Stafford, Cernan and Young on another round trip during which the lunar module was tested, swooping down to only 10 miles from the ground. Finally, in July 1969, Apollo 11 fulfilled the promise that had been made years earlier by President Kennedy. While Michael Collins circled the Moon in the Command Module of the space-craft, Neil Armstrong and Edwin Aldrin stepped out on to the bleakness of the Mare Tranquillitatis. Just before 3 a.m. on July 21, listeners all over the world heard Armstrong's words as he set foot on the Moon: 'That's one small step for a man, one giant leap for mankind.'

The astronauts' stay was relatively brief. Altogether Armstrong was outside the grounded lunar module for 2 hours 47 minutes, Aldrin rather less. During that time they set up various experiments, notably a seismometer or 'moonquake-recorder' which was remarkably sensitive and was designed to measure any minor tremors which might occur in the lunar crust. They had no

DAMAGED SERVICE MODULE OF APOLLO 13, *which was photographed after being finally jettisoned*

THE APENNINES *(upper), photographed from Apollo 15 before it landed.* ALAN SHEPARD ON THE MOON *(lower), Apollo 14, 1971. Note the intensely black shadow cast by the Lunar Module*

difficulty in moving around, despite their cumbersome space-suits, and there was no evidence of any unsafe area around. Blasting back into orbit was accomplished faultlessly; and at 4.49 a.m. on July 24 they splashed down safely in the Pacific after a journey which had lasted for 195 hours. Their time of arrival was precisely 30 seconds late.

Apollo 12 was sent up in November of the same year; this time Charles Conrad and Alan Bean made the descent, landing exactly on target close to the grounded Surveyor 3 automatic probe which had been on the Moon since April 1967, but whose power had long since failed. The astronauts were even able to bring back parts of it, including the camera. After they had set up their experiments, undertaken two moon-walks and blasted back to rejoin their colleague Richard Gordon in the orbiting Command Module, the ascent stage of the lunar module was deliberately crashed back onto the Moon, setting up vibrations which were recorded by the seismometer and which went on for almost an hour.

Few people need to be reminded of the drama of Apollo 13, the space-craft of April 1970, in which the three members of the crew—James Lovell, Fred Haise and Jack Swigert—so nearly met their deaths. Everything went wrong; a crippling explosion during the outward journey wrecked the all-important service module, and it was only through miracles of improvisation by the ground controllers, plus the incredible courage and coolness of the astronauts themselves, that disaster was averted. The one scientific result was again with the seismometer. The final stage of the massive launcher was crashed onto the Moon, and the equipment which had been deposited there by Conrad and Bean recorded vibrations which persisted for three hours.

Unquestionably the setback was severe, but it could be—and was—overcome. Apollo 14, of February 1971, restored the situation. It was perhaps fitting that the commander of the expedition should be Alan Shepard, who had been America's first man in space almost ten years earlier. With him was Edgar Mitchell; Stuart Roosa remained in orbit around the Moon. This time the landing was in the uplands near the old crater Fra Mauro; success was complete. Neither had the Russians been idle. In 1970 they had managed to send an automatic probe, Luna 16, to the Moon and back; it had landed in the Mare Fœcunditatis (Sea of Fertility), and had returned with samples of moon-rock. Even more remarkable was Lunokhod 1, an extraordinary-looking vehicle which had been taken to the Moon in Luna 17 of November 1970, and had crawled about the Mare Imbrium, sending back photographs as well as important data. It continued to operate well into the following year, 'parking' during the long nights and recharging its solar batteries during the equally long periods of daylight.

Quarantining of the astronauts and their lunar samples had been strictly enforced during all the early flights, even though it was generally agreed that the risks of bringing back harmful material were absolutely negligible. The Moon-rock proved to be of intense interest and of essentially basaltic type; it was very old, and it seemed that the Moon and the Earth must be of around

COMMAND AND SERVICE MODULES OF APOLLO 15 *in orbit around the Moon, 1971*

the same age (approximately 4,700 million years) even though they had never formed one body. The lunar magnetic field was extremely weak, and atmosphere was lacking. There was no evidence of hydrated material, which effectively disposed of the theory that the maria had once been water-filled; and, as fully expected, there was a total absence of any trace of life, either past or present. Evidence from the seismometers confirmed that mild 'moonquakes' do occur, and that they are commonest near perigee. Moreover, the Apollo 14 equipment recorded what seemed to be gaseous emissions from below the crust, which came as no real surprise to those who—such as myself—had always believed in a certain amount of activity there.

Then, in 1971, came the epic flight of Apollo 15. Astronauts David Scott and James Irwin landed in the foothills of the Apennines, near the great winding Hadley Rill, and actually went for a drive in a specially-designed 'car', the Moon Rover. Hadley itself proved to be what Scott termed 'a strangely uniform mountain'—very different from the jagged peaks pictured by writers of science fiction. Meantime Astronaut Alfred Worden was orbiting the Moon, carrying out experiments and taking superb photographs. Apollo 15 was the truly scientific mission, and paved the way for Apollo 16 of spring 1972, where the target area was in the rough region of the crater Descartes and once again a Moon Rover was to be carried to the lunar surface (and, incidentally, left there!).

It is true to say that we have learned more about the Moon in the years since 1959 than men had been able to do in the previous two thousand years. No doubt manned bases will be established there well before the end of the century, and will be of tremendous value to humanity. Terrestrial isolation is at an end.

Has the Moon lost any of its magic and its romance? Some people claim so; but I cannot agree. Lifeless and desolate though it may be, it retains its fascination. It is more than our faithful companion in space; it is our stepping-stone to other worlds.

APOLLO 15. *After the landing (right), James Irwin is seen saluting the U.S. flag. The mountain in the background is Hadley Delta. David Scott and James Irwin (above) are in the Lunar Rover and (below) David Scott drills into the lunar surface*

19　the Sun's family

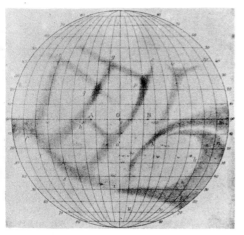

MAP OF MERCURY *by G. V. Schiaparelli*

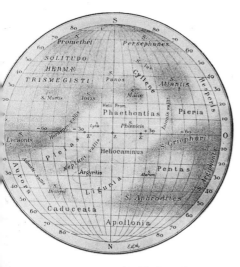

MAP OF MERCURY *by E. M. Antoniadi drawn from observations made with the Meudon 33-inch refractor. Antoniadi's nomenclature is now in general use*

THE SUN HAS AN interesting family. Its senior members are the nine planets, all of which have their own special features. We know a great deal about the third planet in order of distance, since it is the Earth on which we live, but with regard to the rest our knowledge is still far from satisfactory.

Any small modern telescope will show the phases of Venus, the dark patches and white caps of Mars, the moons of Jupiter and the rings of Saturn. With larger telescopes, such as are possessed by many serious amateur observers, it is possible to do really useful work—particularly since the great reflectors at Palomar and elsewhere are kept busy on their studies of distant stars.

First, then, let us look briefly at each planet in turn, beginning with Mercury and working our way outward from the Sun.

The trouble about Mercury is that it is relatively close to the Sun, and never becomes conspicuous. It is always somewhere near the Sun in the sky, and is at its highest during daylight, when it is invisible without a good telescope equipped with setting circles. There is a story that Copernicus never saw Mercury in his life, owing to mists rising from the River Vistula near his home. This is probably 'just another tale'—after all, Copernicus spent some time in Italy, where there is little mist or fog—but we have to agree that Mercury is not easy to find with the naked eye. When visible at all, it appears either low in the west after sunset, or low in the east before sunrise. At its best it looks like a rather pinkish 1st-magnitude star.

Mercury often twinkles. In the ordinary way a planet twinkles much less than a star, because it shows a small disk instead of being a mere point of light; twinkling, of course, is due entirely to the unsteadiness of the Earth's air. Mercury, however, is never visible to the naked eye at all except when near the horizon, so that it is shining through a thicker layer of atmosphere.

Small or even moderate-sized telescopes will show little on Mercury, though the phase is easy enough to see. Schröter drew it often, and also tried to draw a map of the surface markings. He even believed that he had detected a mountain 11 miles high. This was one of his mistakes; there may well be peaks on Mercury, but it is not likely that they reach so great an altitude.

The first reasonably good map was drawn by an Italian astronomer, Giovanni Schiaparelli, more than half a century after Schröter's time. Schiaparelli, born in 1835, became Director of the Brera Observatory, Milan, in 1860, and resigned only when his eyesight began to fail him; it is tragic to record that he was blind for several years before his death in 1910.

Schiaparelli was an excellent observer. He decided to study Mercury during the daytime, with the Sun well above the horizon, since the planet would then be higher up and its image would be steadier. He managed to draw various dark patches on the surface, and gave them names. His observations led him to believe that the axial rotation period must be equal to the revolution period—88

MERCURY, *April 23, 1956. 15 h., 12·5 in. reflector, power 470. Drawing by Patrick Moore*

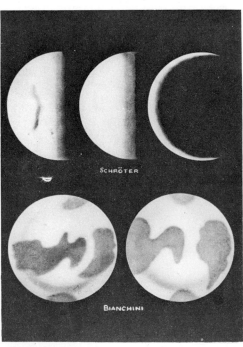

OLD DRAWINGS OF VENUS *by Bianchini and Schröter. The markings recorded by Bianchini, at least, are certainly illusory*

Earth days—in which case Mercury would keep the same face turned toward the Sun all the time. There would be a part of the planet over which the Sun would never set, with a corresponding area of everlasting 'night', and an intermediate or Twilight Zone between these two extremes.

Then, in 1962, W. E. Howard and his colleagues at the University of Michigan, measured the long-wavelength radiations coming from Mercury, and found that the dark side was much warmer than it could possibly be if it were turned permanently away from the Sun. Shortly afterwards radar methods were brought into the investigation by Rolf Dyce and Gordon Pettengill, at Arecibo in Puerto Rico. What they did was to 'bounce' radar pulses off Mercury, and measure the rate of spin. The results were apparently conclusive. Mercury has a rotation period of only 58½ days, so that every part of its surface is sunlit at one time or another. There seems to be some association with the position of the Earth, because every time Mercury is best placed for observation, the same hemisphere is facing us.

Not surprisingly, astronomers of the pre-1960 period were misled; even E. M. Antoniadi, who used the powerful 33-inch refractor at Meudon, near Paris, to draw up what still remains the best map of the surface features, was convinced of the 88-day rotation. Less plausibly, he believed that the planet had an atmosphere dense enough to support dusty clouds. This has not been confirmed in recent times, and it is now thought that Mercury is to all intents and purposes without atmosphere.

Certainly it is a most hostile world. To an observer on the equator, the interval between sunrise and sunset would be equal to 88 Earth-days, and the apparent movements of the Sun would be peculiar, partly because of the slow axial rotation of the planet and partly because the orbit is comparatively eccentric (the distance from the Sun ranges between 43½ million miles and only 28½ million miles). To a Mercurian observer who sees the Sun at perihelion when also at the overhead point, there will be a period when the Sun seems to move 'backwards' for a while near the zenith before resuming its original direction of movement. The maximum temperature must exceed 700 degrees Fahrenheit.

Various maps of Mercury have been drawn in recent years, though it is doubtful whether any of these represent an advance over Antoniadi's chart of the 1930s. The next step will be taken by the fly-by probe scheduled to pass near Mercury in March 1974. If pictures are obtained, we will know whether or not the surface is of lunar type. There may well be craters, but the relief is likely to be less sharp than in the case of the Moon, because the temperature range between day and night on Mercury is even more extreme, and the surface formations will probably be broken by the alternate expansion and contraction. Obviously there can be no hope of finding life there, and a manned expedition will be a very difficult matter indeed.

Venus is very different. It is about the same size as the Earth; indeed, if it is represented by a billiard-ball, the Earth would be another ball so similar to it that the two would have to be carefully weighed to find out which was which. Venus has a diameter of 7,700 miles as against 7,926 for the Earth.

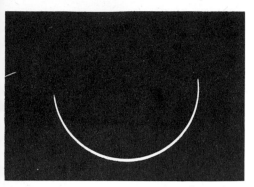

DRAWING OF VENUS, *made on September 1, 1959, by W. M. Baxter with a 4-inch refractor. At this time Venus was practically at inferior conjunction. The curve of the terminator is correctly shown as smooth; the serrations drawn by older observers such as Schröter are due to effects of the Earth's atmosphere, and have nothing directly to do with Venus itself*

TWO PHOTOGRAPHS OF VENUS, *taken with the 100-inch reflector at Mount Wilson*

Like Mercury, Venus shows phases from new to full, since it is closer to the Sun than we are. Yet very little else can be made out. As both Schröter and Herschel realized, Venus is covered with atmosphere, and this atmosphere is so 'cloudy' that we cannot see through it. All we can observe is the top of a cloud-layer, which, incidentally, explains why Venus looks so brilliant.

This means that direct observations, visual or photographic, are painfully limited in scope. Vague, shady features may be seen often enough, but are never definite; occasional brighter areas have also been reported, and during the crescent stage the horns or 'cusps' are generally bright. There have also been many observations of the so-called Ashen Light, or faint luminosity of the dark side of Venus. Yet little was known about Venus as a world, and even the length of the rotation period remained unknown. The American astronomer G. P. Kuiper favoured a period of about four weeks, but there were also suggestions that Venus, like Mercury, might keep the same face turned permanently sunward.

Direct measures showed that the temperature of the uppermost part of the cloud-layer was low, but it was thought that the surface must be hot. As long ago as 1932, W. S. Adams and Theodore Dunham, at Mount Wilson, had analyzed the atmosphere of Venus, using spectroscopic equipment together with the 100-inch reflector, and had detected large amounts of carbon dioxide. Since carbon dioxide tends to act as a 'blanket', shutting in the Sun's heat, a high surface temperature was to be expected. The general view was that Venus was likely to be a barren dust-desert. If so, no life of any sort could be expected there, and in its way Venus would be just as hostile as the airless, waterless Moon.

Opinions changed in 1959, when two Americans, Commander Ross (the pilot) and C. B. Moore, went up in a balloon and took spectrograms of Venus from above much of the Earth's atmosphere, detecting what were regarded as indications of water-vapour. This lent support to a novel theory by F. L. Whipple and D. H. Menzel that Venus might be mainly ocean-covered. The curious Ashen Light was attributed to electrical effects in the planet's atmosphere, and it was thought that Venus must have a strong magnetic field.

Then, in 1961, the Russians dispatched the first Venus Probe, designed to pass close to Venus and send back positive information. Unfortunately the experiment failed, since radio contact with the Probe was lost at an early stage and was never regained, but in the following year the Americans had a brilliant success with their probe Mariner II.

Mariner II was launched on August 27, 1962, from Cape Kennedy (then known by its old name of Cape Canaveral). Originally it was planned to pass over the sunlit side of Venus at a distance of less than 10,000 miles; its distance from Earth would then have been well over 30,000,000 miles. Slight errors in launching and guidance meant that the closest approach to Venus was over 20,000 miles, but even this represented great accuracy. Moreover, when the rocket and the planet were at their nearest, on December 14, radio contact was good.

When the results were announced, in 1963, astronomers in

general were greatly surprised. Almost all the Mariner information proved to be unexpected. Venus appeared to have no magnetic field; the surface temperature was over 800 degrees Fahrenheit, and the rotation period was very slow. Almost all preconceived ideas about Venus had to be jettisoned—and, in particular, the attractive ocean theory put forward by Whipple and Menzel had to be cast aside. Liquid water could never exist at such a temperature, even allowing for high atmospheric pressure.

Various more recent probes have refined and confirmed the Mariner II findings. One of these (Mariner V, which by-passed Venus within 2,500 miles in October 1967) has been American, but the rest have come from the U.S.S.R., and the Russians have continued to show strong interest in Venus. Their vehicle Zond 1 of 1964 was a failure inasmuch as contact with it was lost, and the same fate overtook Venera 2 and Venera 3, which were launched in the following year—though Venera 3 seems to have crash-landed on the planet on March 1 1966. Then in 1967, came Venera 4, which made a controlled parachute descent through the atmosphere of Venus. The capsule containing the instruments went on transmitting until it was only around 20 miles above the ground. At this point the crushing atmospheric pressure and the high temperature put it out of action, though the situation did not become clear for some time.

In January 1969 Veneras 5 and 6 were sent up, and their instrument capsules soft-landed on the planet on May 16 and 17 respectively. Again the signals failed before impact, and it was not until 1970 that complete success was achieved. Venera 7, launched on August 17, landed on December 15, and signals from it were received for another half-hour. At that moment Venus was some 37,000,000 miles from the Earth. This meant that a radio signal took 3 minutes 22 seconds to travel from one world to the other.

Summing up the results from these various probes, we can now make some definite statements about conditions on Venus. First, the atmosphere is extremely dense. The ground pressure is between 80 and 110 times that of our own air at sea-level. The main constituent is carbon dioxide, which makes up approximately 97 per cent of the total. This means that there is not much room left for anything else; nitrogen, oxygen and water vapor are in very short supply. With reluctance, we must abandon any hope of finding advanced life-forms on Venus. Neither will journeys there be easy; the surface temperature has been measured as about 890 degrees Fahrenheit.

The very slow rotation has been confirmed by radar measurements carried out from Earth. The official value is now 243 to 244 days, which is longer than the planet's 'year' of $244\frac{3}{4}$ days. Again there is a 'locking' effect with respect to the Earth, which may or may not be due to coincidence. Venus rotates exactly four times between successive inferior conjunctions, so that each inferior conjunction the same hemisphere is facing us. To an observer on the planet itself, the Sun would rise in the west and set in the east 117 Earth-days later; but needless to say, neither it nor the Earth could ever be seen through the dense carbon-dioxide atmosphere. Venus must be an eerie world. From the surface,

APPARENT SIZE OF VENUS, *from new to full, shown to the same relative scale*

PHOTOGRAPH OF VENUS, *taken on February 19, 1961, by H. E. Dall with his $15\frac{1}{4}$-inch reflector*

VENUS. *From observations by Patrick Moore. (Upper) July 17, 1959, 16h. 55m., 24-inch reflector × 350. (Lower) January 3, 1958, 16h. 30m., 12·5-inch reflector × 250. The Ashen Light is shown, but has been somewhat exaggerated for the sake of clarity. Drawing by D. A. Hardy*

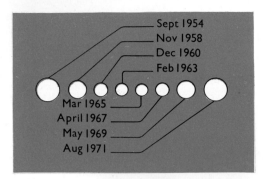

APPARENT OPPOSITION SIZE OF
MARS. *The maximum size for each opposition between 1954 and 1971 is shown to the same scale*

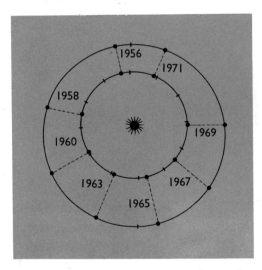

OPPOSITIONS OF MARS, 1956–71. *It will be seen that the most unfavorable oppositions are those of 1963 and 1965*

light-rays would be 'super-refracted', and it has been suggested that an observer would have the impression of being at the bottom of a large bowl, with the horizon curving upward on every side!

We still know little about the nature of the surface itself. Radar measurements carried out in America by a research team headed by R. M. Goldstein and H. Rumsey have led to a map of part of the surface, showing several roundish features which may or may not be elevated plateaux; one of these, provisionally called Alpha, is 600 miles across. More reliable data can hardly be secured until the time comes to put an automatic vehicle into orbit around Venus. No magnetic field has been detected as yet, and the cause of the Ashen Light remains a mystery.

Why should Venus, so nearly the Earth's twin in size and mass, be so different in state? We have to admit that we do not know, though the lesser distance from the Sun (67,000,000 miles, as against our 93,000,000 miles) must surely have something to do with it. More probes are being planned, mainly by the Russians, but a manned expedition to the surface seems very unlikely in the foreseeable future. Instead of being comparatively welcoming, Venus has proved to be even more hostile, in its own way, than Mercury or the Moon.

In one further respect Venus differs from the Earth: it has no large natural satellite. This now seems certain, though it was formerly thought that a satellite existed. The first serious observations of it were made in August 1686 by G. D. Cassini, in Paris, who reported a body with a diameter about one-quarter that of Venus, showing the same phase. Further accounts of it were given with fair regularity during the next seventy years, notably in 1740, when the satellite was seen by the English telescope-maker James Short. Then, in 1761, Montaigne of Limoges recorded the satellite several times, and for a brief period it existence was regarded as established. An orbit was calculated by the German mathematician Lambert, who found that the distance from Venus was 259,000 miles, and the period 11 days 5 hours. If the estimates by Montaigne and others were correct, the diameter would be in the region of 2,000 miles, much the same as that of the Moon.

The satellite was again observed in 1764 by three astronomers, two in Denmark and one in France, but these were the final reports. Subsequently the satellite disappeared from the record-books, and despite careful searches by Herschel and others it has never been recovered. Consequently, there is no escape from the conclusion that it never existed at all. Venus itself is a brilliant object, and the satellite must be dismissed as a 'telescopic ghost'.

Of course, it is quite possible that Venus possesses a very small satellite, but such a body can hardly be more than a mile or so in diameter, inferior even to the two dwarf attendants of Mars. For the moment, it is reasonable to conclude that Venus, like Mercury, is a solitary wanderer in space.

Mars, the first planet beyond the orbit of the Earth, is completely different from Venus. In size and mass it is intermediate between the Earth and the Moon; its diameter is 4,200 miles, and its mass is about one-tenth that of our world, giving an escape velocity of 3.1 miles per second. At its closest to us, as in August 1971, it can come within 35,000,000 miles, and is then brighter than any object in the sky apart from the Sun, the Moon and Venus. It comes to opposition only every alternate year (1971, 1973, 1975 and so on) but is always readily identifiable because of its strong red color. It is hardly surprising that the ancients named it in honor of the God of War.

Mars has a somewhat eccentric orbit; its distance from the Sun ranges between $129\frac{1}{2}$ million miles at perihelion out to 154 million miles at aphelion. Obviously, the most favorable opportunities for study take place when opposition occurs near perihelion, as in 1971. At such times even a small telescope will show definite markings on the disk. The dominant color is reddish-ochre, but there are dark patches which are unquestionably permanent; they were first recorded by Huygens as long ago as 1659. Moreover, the poles are covered with whitish caps which show a seasonal cycle, reaching their maximum extent in Martian winter and becoming very small in summer; the southern cap has been known to vanish completely. The caps, too, have been known for a long time. The first observation of them was made by G. D. Cassini in 1666.

As with Venus, so with Mars: the information sent back by space-probes during the past few years has caused a complete change in our views, and we have to admit that Mars is not nearly so welcoming as we used to think. Maps have never been a problem, because the thin atmosphere is transparent, and we can see straight through to the surface; as early as 1877 the Italian astronomer G. V. Schiaparelli drew a chart which was accurate enough to be really useful, and there is no difficulty whatsoever in mapping the main features. To Schiaparelli we owe the modern nomenclature; the dark areas have names such as the Syrtis Major, Mare Tyrrhenum, Mare Erythræum and so on. They are less romantic than the original names of the pre-1877 period; thus the somewhat triangular Syrtis Major used to be called the Kaïser Sea or the Hourglass Sea. However, the Schiaparelli system will certainly not be altered now.

It was natural to suppose that the ochre tracts were deserts, the dark regions seas, and the polar caps snow or ice-fields, but well before the end of the last century it had become clear that the Martian atmosphere is not only very thin but also very dry. Large

MARS AND THE PLEIADES. *September 16, 1952, 1·05 hours. Photograph by Ramon Lane*

MAP OF MARS

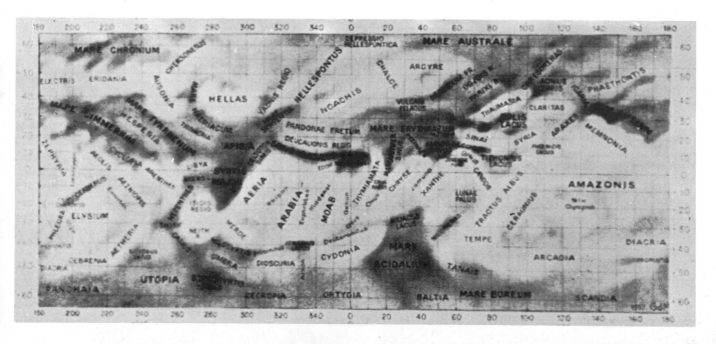

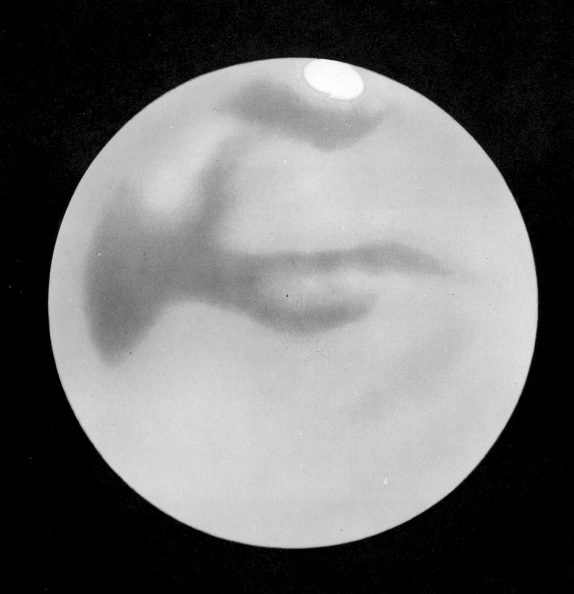

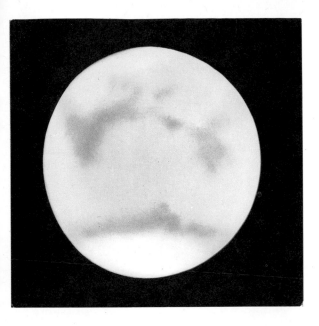
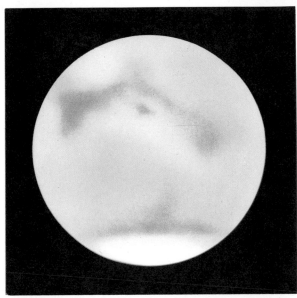
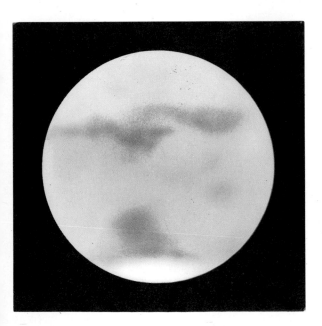
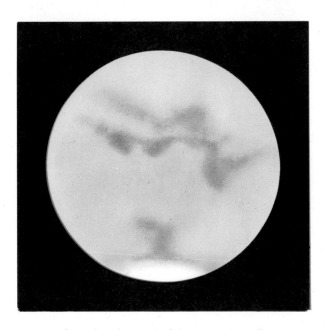

SEASONAL DECREASE OF THE NORTH POLAR CAP OF MARS. *From observations by Patrick Moore.* (Top left) *November 19, 1960, 3h. 10m., 12·5-inch reflector × 450— note the phase, since Mars was some way from opposition.* (Top right) *December 25, 1960, 0h. 20m., 8·5-inch reflector × 400.* (Lower left) *January 11, 1961, 17h. 55m., 8·5-inch reflector × 300.* (Lower right) *February 6, 1961, 22h. 20m., 24-inch reflector × 525. Drawings by D. A. Hardy*

tracts of water were found to be out of the question. Instead, most astronomers followed a theory of E. Liais', also in 1877, to the effect that the dark regions were covered with vegetation. It was not thought that this vegetation could be of advanced type; trees, bushes and flowering plants could hardly survive on Mars, where the atmosphere is so tenuous and the night temperature is so low (more than 100 degrees below zero Fahrenheit, though at noon on the equator a thermometer might rise to almost 70 degrees Fahrenheit). More plausibly it could be the Martian equivalent of moss or lichen—though it would be most unwise to draw any direct analogy with terrestrial vegetation.

Strong support for the vegetation hypothesis came from the seasonal cycle. When the polar caps shrank, what was often termed a 'wave of darkening' spread equatorward, affecting first the dark regions in high latitudes and extending to the temperate and equatorial zones. Vegetation revived from winter hibernation

MARS. *From an observation by Patrick Moore. October 12, 1956, 21h. 10m., 12·5-inch*

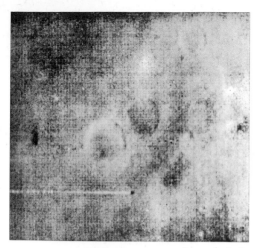

THE CRATERS OF MARS, *photographed from Mariner IV. This is Frame No. 8*

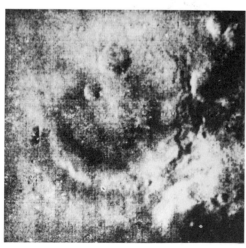

THE CRATERS OF MARS, *photographed from Mariner IV. This is the 11th Frame— the best obtained from Mariner*

by the arrival of moisture-laden breezes wafted from the caps, would be expected to behave in such a manner. The ochre tracts were thought to be covered with some colored mineral such as felsite or limonite, and it was agreed that the polar caps themselves must be very thin, with a depth of no more than an inch or two at most.

Tentative measurements of the atmosphere gave an estimated ground pressure of 85 millibars, which is equivalent to the pressure in our own air at a height of rather over 50,000 feet above sea-level. This would be too thin for Earth-type creatures to breathe even if it were made up of pure oxygen, but oxygen was obviously in short supply, and it was believed that the atmosphere must be composed chiefly of nitrogen, together with a little free oxygen and water vapor and a slightly higher percentage of carbon dioxide. The clouds seen on frequent occasions were of the high, fleecy variety (ice crystals?) or the lower, yellow veils (dust-storms). It was also assumed that the atmosphere provided an adequate shield against both meteors and potentially dangerous short-wave radiations sent out by the Sun.

Mountains were not expected, and Mars was regarded as a somewhat flat landscape. The dark regions were known to be warmer than the deserts, and were taken to be depressed areas, possibly old sea-beds. All in all, there seemed no reason to doubt that Mars would be a practicable site for an elaborate manned base in the future. Day and night conditions would be familiar, as the planet has a rotation period of 24 hours 37 minutes; and the seasons would be of the same type as ours, though they would be regulated by the much longer Martian 'year' of 687 Earth-days.

Then, in 1964, came the launching of America's probe Mariner IV, which by-passed Mars at a mere 6,300 miles on July 15 of the following year. Photographs taken with its cameras showed that far from being smooth, the Martian surface is cratered in much the same way as that of the Moon. More significantly, it was found that the atmosphere is much less dense than had been previously thought. Full confirmation was obtained with the next two successful Mars probes, Mariners VI and VII, which flew by the planet at little over 2,000 miles in the late summer of 1969 (July 31 and August 5 respectively). It now seems that the ground atmospheric pressure is no more than 7 millibars, perhaps only 6, and that the main constituent is carbon dioxide. As a screen against short-wave radiations, the atmosphere is likely to be of little use.

Though final conclusions must be deferred until we have additional information, most astronomers now think that the polar caps are due to solid carbon dioxide rather than to ice or frost—and that the dark regions are not vegetation-covered at all. Shifting dust may be a more plausible answer, though admittedly the seasonal cycle remains a puzzle. Neither are the dark regions low-lying; some of them, such as the Syrtis Major, are elevated, as has been shown by measurements carried out by R. A. Wells with regard to the carbon dioxide pressure over various areas of the surface.

Mariners IV, VI and VII between them sent back photographs of only about 20 per cent of the Martian surface, but they proved that the terrain is not the same everywhere. There are cratered

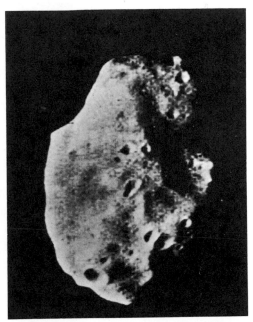

PHOBOS, *photographed from Mariner IX in 1971*

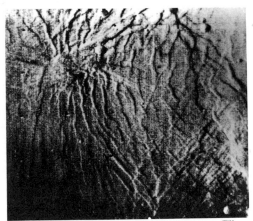

SURFACE FAULTS ON MARS. *The Phœnicis Lacus plateau area from Mariner IX during its 67th orbit*

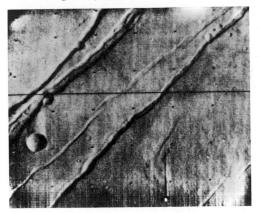

RILLS ON MARS, *in the Mare Sirenum, from Mariner IX*

regions; chaotic areas, without craters but with tangled mountains, pits and valleys; and one circular plain, known as Hellas (one of Schiaparelli's names), is featureless, so that presumably any craterlets which used to exist inside it have been melted down or eroded away. Some of the craters are huge, and measure up to 300 miles in diameter. One of them has been known for many years, and has been named Nix Olympica. I have often seen it with the 12½-inch reflector in my own observatory, but it looks like nothing more than a tiny speck, and nobody had any idea of its true nature until the Mariner pictures were received.

In May 1971 the Americans launched two more probes, Mariners VIII and IX. The first of these failed immediately after being dispatched, and fell unceremoniously into the sea, but Mariner IX went on its way toward Mars, and reached the neighborhood of the planet in the following November. It was then put into an orbit round Mars itself—to be greeted by a dust-storm which obscured vast areas of the planet's surface! As soon as the dust cleared, superb pictures began to be received, showing large structures which were rather clearly volcanic calderas, together with shallow rills and strange features which looked superficially like riverbeds. The sharpness of the features suggested that they were geologically young—and today active vulcanism on Mars cannot be ruled out. Also, Mariner IX photographed the two dwarf satellites, Phobos and Deimos. Both proved to be irregular in shape, and pitted with craters. Phobos is an extraordinary body, 16 miles long by 13 wide, giving every impression of being a 'bit of something'; Deimos is of the same basic type.

Meantime, two Russian probes, Mars 2 and Mars 3, had joined Mariner IX in orbit. The first deposited a Soviet pennant on the planet, which may have been politically impressive but was not actually of much use. Mars 3, however, soft-landed a capsule in the area between Electris and Phæthontis. The landing procedure worked perfectly—a magnificent triumph for Russian technology. Alas, after transmitting from the surface for 20 seconds the signals failed, and contact was never re-established. It seems that we must wait until 1976, when America's Viking probe will soft-land, before we can finally decide whether there is or is not life on Mars.

Phobos and Deimos had been discovered in 1877 by Asaph Hall at the Washington Observatory. From the first they were known to be strange bodies. Both are close to Mars, and Phobos has a revolution period of only 7h. 39m.—shorter than the Martian 'day' of 24h. 37m. They are quite different from our own massive Moon, and the Mariner photographs suggest that they may be nothing more than asteroids which were captured by Mars in the remote past.

No description of Mars would be complete without mention of the 'canals'—straight, artificial-looking streaks on the surface first described in detail by Schiaparelli in 1877. Percival Lowell, an American ex-diplomat who turned to astronomy, founded the Flagstaff Observatory in Arizona mainly to observe them; he firmly believed them to be artificial, making up a planet-wide irrigation system. His ideas caused fierce argument even during his lifetime (he died in 1916), but long before the space age the idea of intelligent Martians had been discounted. It now seems

1951

1956

1959

1960

THE CHANGING ASPECTS OF JUPI-
TER'S BELTS. *From observations made by
Patrick Moore.* (1) *September 1, 1951, 0h.
19m., 8·5 inch reflector × 350.* (2) *April 17,
1956, 21h. 35m., 12·5-inch reflector × 350.*
(3) *March 28, 1959, 3h. 33m., 12·5-inch
reflector × 360.* (4) *August 1, 1960, 20h.
58m. 12·5-inch reflector × 300. Note the
changes in the north and south equatorial belts,
and the unusual coloration of the equatorial
zone in 1959. Drawing by D. A. Hardy*

*JUPITER. From an observation by Patrick
Moore, June 5, 1958, 21h. 52m. 12·5-inch
reflector × 400. Drawing by D. A. Hardy*

that some of the canals have a basis of reality inasmuch as they are
either raised mountainous areas or else chains of craters, but they
are undoubtedly natural features.

In every way Mars is a fascinating world. Admittedly it is less
welcoming than we believed before the Mariners flew past it, but
it remains more Earthlike than any other planet, and we cannot
yet be sure that it is sterile. The first astronauts who go there—
perhaps about 1990?—may have many surprises in store for them.

Beyond Mars comes the zone of the asteroids, and still farther out
we reach the four giants—Jupiter, Saturn, Uranus and Neptune.
All are huge, and all seem to be built upon the same pattern, so
that they have many points in common.

A century ago it was thought possible that the giants might be
'dwarf suns' with hot surfaces, sending out considerable light
and heat of their own. We now know that this is not the case, and
that all four are bitterly cold, with temperatures ranging from
−200 degrees Fahrenheit for Jupiter down to −360 degrees for
Neptune.

The outer gases can be analyzed by means of the spectroscope,
and it is found that most of the outer 'clouds' are made of hydrogen
together with hydrogen compounds such as ammonia and methane
(marsh-gas). According to a theory due to Rupert Wildt, of the
United States, a giant planet consists of a rocky core surrounded
by a deep layer of ice which is in turn overlaid by the hydrogen
atmosphere; with Jupiter, the core would be 37,000 miles in
diameter, the ice-layer 17,000 miles thick and the atmosphere
8,000 miles thick. Another idea, due to W. H. Ramsey of Britain,
suggests that a giant is composed mainly of hydrogen all the way
through, though near the centre of the globe the hydrogen will be
under such tremendous pressure that it will start to behave like a
metal. And according to a still more recent theory by P. Peebles,
there is a core of dense metals and rocky silicates, over which
come, in order, a layer of metallic hydrogen, a layer of more ordinary solid or liquid hydrogen, and the outer atmosphere. Almost
certainly Jupiter is hot inside, with a core temperature of half a
million degrees.

Jupiter spins on its axis very quickly. The rotation period is only
$9\frac{3}{4}$ hours, and this means that the equator bulges out—giving the
planet a flattened shape, as shown in the drawings. Different
regions rotate at different speeds, and the region near the equator
(System I) has a period several minutes shorter than for the rest of
the planet (System II). Moreover, special features, such as spots,
have rotation periods of their own.

Through a small telescope Jupiter shows as a yellowish, flattened disk, crossed by the streaks which we term *belts*. Generally
a 3-inch refractor will show at least four belts, and sometimes as
many as six, while with larger instruments there is a tremendous
amount of detail. As Jupiter rotates, the markings are carried
from one side of the disk to the other—as with sunspots, though
here the time taken to pass from limb to limb is only five hours
instead of a fortnight. The shifting becomes noticeable after only a
few minutes. By timing the moments when certain features cross
the central meridian, it is possible for astronomers to measure the
rotation periods of the various zones with great accuracy.

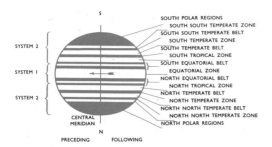

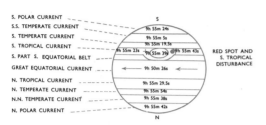

ROTATION PERIODS OF THE VARIOUS ZONES OF JUPITER. *The 9h. 50m. 26sec. zone is termed System I; the rest of the planet, System II*

JUPITER, *photographed with the 60-inch reflector at Mount Wilson.*

SATURN, *photographed with the 200-inch Hale reflector at Palomar*

Much of this work has been carried out by amateurs, and in particular by members of the Jupiter Section of the British Astronomical Association. This Association was formed in 1890, and has a distinguished record of observational work; indeed, its published Reports show that Jupiter has been thoroughly studied for more than 70 years now. It is probably true to say that Jupiter is the best of all objects for the serious observer armed with a small or moderate telescope.

The details in the belts are always changing, and this applies also to the spots. Special features on Jupiter do not generally last for long, as is only to be expected, since the surface is made up of gas. The chief exception is the remarkable object known as the Great Red Spot.

The Spot first became prominent in 1878, developing from a pale pink, oval marking into a brick-red area 30,000 miles long by 7,000 miles wide, so that its surface area was equal to that of the Earth. It attracted a great deal of attention, and was traced with fair certainty on earlier drawings. Schwabe, discoverer of the eleven-year sunspot cycle, had recorded it in 1831, and Hooke may have seen it as long ago as 1664.

The bright red hue did not persist in so violent a form, and since 1890 the Spot has shown variations in both hue and visibility. It has even been known to vanish at times, but it has always returned and during the period from the mid-1960s to the present (1972) it has been striking; I have recorded it as salmon-pink at times. It does not move much in latitude, but it drifts along in longitude, so that its rotation period is often slightly different from that of other features in the same area.

What precisely is the Red Spot? We must confess that we do not know. Certainly it is not the top of a volcano poking out through the cloud-layer, as used to be thought. There are two theories both of which have their supporters—and both of which may prove to be wrong. The Spot may be a solid or semi-solid mass floating in Jupiter's upper atmosphere, or it may be the top of a pillar of stagnant gas known as a Taylor column. At any rate it is unique, and must be more significant than a mere 'cloud'.

Jupiter's four chief satellites, first seen in 1609 by Marius and Galileo, are easy objects; some exceptionally keen-eyed people can see them without a telescope. Eight more moons have been found but all are tiny and require powerful instruments. No. 5, discovered by E. E. Barnard in 1892, is closer to the planet than any of the 'Galileans', and goes round once in 11 hours 57 minutes.

Two of the main satellites, Ganymede and Callisto, are larger than the planet Mercury, but seem to be without atmosphere. The remaining two, Io and Europa, are much the same size as our Moon. Astronomers at the Pic du Midi have recorded surface features on all four, and have even gone so far as to draw up preliminary maps.

It is hardly necessary to add that no landing on Jupiter will ever be possible, but in the future it should be practicable to go to some of the satellites. Meanwhile, automatic probes are being planned and the first of these will—if there are no hitches!—by-pass the planet toward the end of 1973. As well as taking photographs, and making measurements of temperatures and atmospheric const

SATURN IN 1933, *showing the famous white spot discovered by W. T. Hay. This is Hay's original drawing*

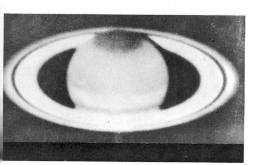

SATURN, *photographed at the Lowell Observatory*

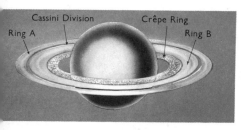

DIAGRAM OF SATURN'S RING SYSTEM *showing the Cassini Division, Rings A and B, and the Crêpe Ring (Ring C)*

tuents, it may help to solve some of the puzzles concerning the radio waves from Jupiter which seem to indicate the presence of a powerful magnetic field there.

Saturn, the outermost of the planets known in ancient times, is much more remote than Jupiter. Its average distance from the Sun is 886 million miles, and its 'year' is over 29 times as long as ours. Moreover it is smaller than Jupiter, and even more flattened at the poles. The diameter as measured through the equator is just over 75,000 miles, while the polar diameter is less than 70,000 miles.

The belts are much less conspicuous than Jupiter's, though a moderate telescope will show several. Spots are rare, but when they do appear they are of special interest. The most important spot of recent years was found on August 3, 1933, by a British amateur named W. T. Hay.

William Thompson Hay was born at Aberdeen in 1888. He was apprenticed to an engineer, but at the age of twenty-one he gave up this career and went on the stage. Before long he had made a great reputation as a variety artist; many people will remember 'Will Hay', the comic schoolmaster. Later he played leading rôles in film comedies, some of which were written specially for him.

He was interested in astronomy from an early age, and was a skilful observer. The spot on Saturn was discovered with the 6-inch reflector at his private observatory at Norbury, near London, but in fact Hay's most important work was in connection with comets, though he was also a clever instrument-maker.

The 1933 spot became prominent for a time, and was strikingly white. It gradually lengthened, and the portion of the disk following it darkened, as though material were being thrown up from below the surface of the gas-layer. Like other spots on Saturn its period of visibility was short, and nothing similar has been seen since, though a moderately conspicuous spot appeared in 1960.

The glory of Saturn is in its ring-system, which may be well seen with a small telescope when the rings are at their most 'open'. So far as we know there is nothing else like them in the heavens.

The system is of vast extent. From side to side it measures almost 170,000 miles. The outermost ring (A) is 10,000 miles wide; then comes Cassini's Division, with a width of 1,700 miles, and then the brightest ring (B) 16,000 miles wide. Johann Encke, once Director of Berlin Observatory and the man who instructed Galle and d'Arrest to hunt for Neptune, discovered a much less clear-cut division in Ring A. Other minor divisions in the rings, reported from time to time, seem to be dubious.

The inner ring, C—the so-called Crêpe Ring—has a curious history, which takes us back to the early days of astronomy in America.

In 1835 William Cranch Bond, a watch-maker who had taken up astronomy as a hobby, was watching Halley's Comet from his private observatory, together with his son George. At that time there were no large telescopes in the U.S.A., and it occurred to Bond and others that the time had come to remedy matters. Citizens in Boston and elsewhere subscribed the money for a large instrument, and in 1847 a 15-inch refractor was set up in a wooden tube at Cambridge, Massachusetts. This was the start of the now world-famous Harvard College Observatory.

SATURN IN 1949. *From an observation made by Patrick Moore on June 2, 1949, 22 h., 6-inch refractor × 400. The rings were almost closed. Four satellites are shown; from left to right, Dione, Enceladus, Tethys and Rhea. When the ring-system is at this angle to the Earth, surface features on the globe of Saturn are best placed for observation, since when the ring-system appears more open, an appreciable part of the globe is masked. However, the belts are always very much less conspicuous than those of Jupiter. Drawing by D. A. Hardy*

On July 16, 1850 George Bond was observing Saturn with the 15-inch when he noticed a third ring, closer to the planet than the old ones and much fainter. Indeed, it seemed to be almost transparent. The discovery, as interesting as it was unexpected, was made independently at about the same time by a British clergyman, the Rev. William Rutter Dawes.

The 'Crêpe Ring' is not a particularly difficult object to observe, and it is strange that nobody recognized it before Bond and Dawes did so in 1850. Even Herschel, who had discovered the two tiny inner satellites Mimas and Enceladus, had not identified it. There have been suggestions that the ring has brightened up since Herschel's day, but this would be hard to account for, and we must assume that the older observers simply overlooked it. It is worth adding that at various times since 1909 there have been reports of another 'crêpe ring', this time outside the main ones, but it has not been confirmed, and probably does not exist. I made a special search for it with the Meudon 33-inch refractor at various times between 1952 and 1958, but without success. Much more recently P. Guèrin has reported another dusky ring, this time inside the Crêpe Ring, and extending almost to the surface of the planet. It has been claimed that photographs show it unmistakably, but I admit to being somewhat sceptical as yet.

What can the rings be? In 1848 a solution was provided by the French mathematician Édouard Roche, who proved that the system is too close to Saturn to be a solid or liquid sheet. Within a certain limit, now known as 'Roche's Limit', a solid or liquid body would be disrupted by the gravitational pull of the planet—and the rings lie well inside the danger zone. Roche therefore suggested that the rings must be made up of numerous small pieces of material, moving round Saturn in the manner of tiny moons. This idea was proved by the work of an American astronomer, James Keeler, who showed that the inner rings go round faster than the outer ones.

SATURN, *photographed with the 60-inch reflector at Mount Wilson*

VARIOUS ASPECTS OF SATURN'S RINGS. *Drawings by D. A. Hardy*

We are still not certain about the nature of the ring-particles, but according to G. P. Kuiper they are made up largely of ice—or are at least coated with icy material.

Altogether Saturn has ten satellites. Titan, discovered by Huygens in 1655, is the largest satellite in the Solar System, with a diameter of about 3,500 miles. It is also the only satellite known to have an atmosphere. Of the rest, Cassini's four (Iapetus, Rhea, Dione and Tethys) have diameters ranging from 800 to about 1,600 miles, while Mimas and Enceladus are about equal at 300 to 400 miles, and Hyperion, discovered by Bond in 1848, is smaller still. The outermost satellite, Phœbe, found by W. H. Pickering in 1898, is a real dwarf only 150 miles across. It is over 8 million miles from Saturn, and goes round the 'wrong way', in the manner of a car going the wrong way in a roundabout. It is not unique; several of Jupiter's minor moons, all of Uranus', and one of Neptune's do the same.

In December 1966, when the ring-system was edge-on, Audouin Dollfus at the Pic du Midi discovered a new inner satellite, now named Janus. It is not far from the edge of the rings, and is detectable only at the time of edgewise presentation, so that it will not be seen again until 1980; moreover it is small, probably inferior to Mimas. Ironically I saw it several times in the autumn of 1966 without recognizing it as new—so that the sole merit of my observations is that they showed Janus to be visible with a 10-inch refractor!

In 1904 W. H. Pickering had announced the discovery of another satellite, moving in an orbit between those of Titan and Hyperion. He named it Themis, and it was regarded as definitely real; but it has never been seen since, and in all probability Pickering was mistaken.

If Jupiter is remote and inaccessible, Saturn is even more so. Eventually there may be landings on Titan, even though its tenuous, methane atmosphere will be of no direct use to astronauts. Meantime, there is the exciting prospect of a fly-by probe in the late summer of 1980. This vehicle will approach Saturn after having made a comparatively near encounter with Jupiter in the previous year; after it has passed Saturn it will go on to Pluto, which should be reached at the end of 1985. The concept of sending a probe from one planet to another is often called the 'Grand Tour', and is possible at the end of the present decade only because the outer planets are suitably positioned. If the Jupiter–Saturn–Pluto vehicle is not launched in 1977, we must wait for over a century for another lining-up. This does not mean that no probes to Saturn or Pluto can be dispatched in the interim, but the gravitational field of Jupiter will not help, and so the journeys will take longer. The American authorities have announced that for financial reasons they will not attempt the experiment, but the Russians will almost certainly do so. Meanwhile, in March 1972, the Americans launched Pioneer 10, destined to fly by Jupiter, and eventually leave the Solar System altogether.

The two outer giants, Uranus and Neptune, are almost perfect twins. Like Jupiter and Saturn they are gaseous, at least in their outer layers, but their surfaces seem to be less active, even though details are extremely hard to see even with giant telescopes. Uranus

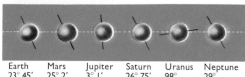

INCLINATION OF THE ROTATION AXES OF THE PLANETS. *The inclinations of Mercury, Venus and Pluto are not known with any certainty. The unusual tilt of Uranus is very evident*

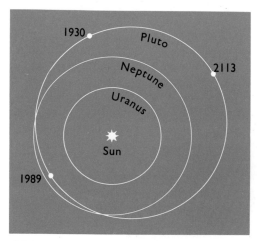

ORBITS OF THE OUTERMOST PLANETS. *Pluto's minimum distance from the Sun is less than that of Neptune, and the planet will not next reach perihelion until 1989. However, there is virtually no danger of a collision, since Pluto's orbit is inclined to the ecliptic at the relatively large angle of 17 degrees*

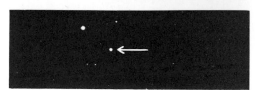

PLUTO, SHOWING APPARENT MOTION OVER 24 HOURS. *Photographed with the 200-inch Hale reflector at Palomar*

orbits the Sun once in 84 years, while Neptune requires $164\frac{3}{4}$ years.

Uranus is interesting because its axis is sharply tilted to the plane of its orbit. The inclination is more than 90 degrees, and the 'seasons' there must be very odd. First much of the northern hemisphere, then much of the southern, will have a night lasting for 21 Earth-years. As seen from Earth, it sometimes happens that we look straight 'up' at the pole of Uranus, as was the case in 1945.

Both the outer giants have satellites. Uranus has five, Neptune two. Of Neptune's attendants, one (Triton) is over 3,000 miles across; the other (Nereid) is a dwarf, and has a very eccentric orbit, so that its distance from Neptune varies between 1 and 6 million miles.

Finally we come to Pluto, tracked down in 1930 by Clyde Tombaugh as a result of Lowell's calculations. It can be seen with a moderate telescope, but looks like nothing more than a faint speck of light, and about its surface features we know nothing at all, though from brilliancy variations the rotation period has been estimated at 6 days 9 hours.

One of the most curious facts about Pluto is that it has an orbit which is unusually eccentric and inclined. At perihelion it approaches us closer than does Neptune, and the diagram shows that it actually 'crosses' Neptune's path, though the tilt of 17 degrees means that there is little fear of a collision. It is, however, possible that Pluto used to be a satellite of Neptune which was disturbed in some way many millions of years ago and has moved off on its own.

The diameter of Pluto has been measured, though with considerable difficulty. The result is decidedly surprising. Apparently Pluto has a diameter of only 3,700 miles, which is smaller than Mars and not very much larger than Mercury. If so, and assuming normal density, it could not produce any measurable effects upon the movements of giants such as Uranus and Neptune. Yet it was by just these effects that its position was worked out by Lowell, with the result that Clyde Tombaugh was able to locate it.

Can the discovery have been due to sheer chance? It seems unlikely but it is equally unlikely that Pluto is super-dense or that our estimates of its diameter are hopelessly in error. The mystery remains, and we can only hope that the Grand Tour probes of the next decade will be able to provide a solution. It is always possible that another even more remote planet awaits discovery, but if such a world exists it is certain to be extremely faint, so that its detection will be largely a matter of luck.

From Pluto, the Sun would appear only as a small though intensely brilliant source of light. Powerful telescopes might show Neptune and Uranus, but the Earth and the inner planets would be quite beyond the range of vision, and our imaginary Plutonian astronomer could never learn of their existence.

We must not over-estimate the importance of the planets. All of them, including the Earth, are very junior members of the Galaxy, and even Jupiter, which we regard as a giant, is insignificant in the pattern of the universe. Yet we are naturally concerned with our near neighbors in space, and the astronomers who have spent countless hours in studying them have not wasted their time.

20 wanderers in space

CUNNINGHAM'S COMET OF 1940, *photographed on December 21 with a 5-inch Ross lens at Palomar*

WE HAVE ALREADY discussed the planet-hunters. Now let us turn to the comet-hunters, starting with the first and perhaps the greatest of them all—the French observer Charles Messier.

Messier was born in Lorraine in 1730, and spent most of his active life in Paris, where he observed with a small telescope installed on top of a tower. He independently discovered Halley's Comet at its return to perihelion in 1759, though he did not see it until after Palitzsch had done so. From that time onward Messier undertook a systematic search for new comets, and discovered twenty-one of them, as well as studying many more which had been found by other astronomers. The French King called him 'the ferret of comets', and the nickname was a good one.

It was during these searches that Messier drew up his famous catalogue of star-clusters and nebulæ. However, comets were his chief concern, and he continued working away until only a few years before his death at the age of eighty-seven.

Caroline Herschel discovered half a dozen comets, and another early 'hunter' was Pierre Méchain, who also worked in Paris and who found eight comets between 1781 and 1799. But the true successor to Charles Messier was Jean Louis Pons.

Pons, born in 1761, went to Marseilles Observatory in 1789—but not as observer or director; he was given the job of doorman and general caretaker. His interest in astronomy was so keen that the observers gave him all the help they could, and Pons later outshone them all. He was made assistant astronomer in 1813, and in 1819 left Marseilles to become Director of the new Observatory at Marlia, near Lucca. He ended his career as Director of the Museum Observatory in Florence, where he died in 1831.

In its way, Pons' career is as romantic as any in science. There can be no other case of an observatory doorman rising to the rank of observatory director, earning an international reputation during his lifetime.

Pons discovered thirty-six comets. One of them, seen in 1818, proved to be of particular interest, because when its orbit was computed by Johann Encke, at that time assistant at Göttingen Observatory, it was found to move round the Sun in a period of only 3.3 years. Working backwards, much as Halley had done, Encke decided that the comet was identical with those observed in 1786 by Méchain, 1792 by Caroline Herschel, and 1805 by Pons himself. In fact, Pons had discovered the same comet twice! Encke predicted that the comet would return in 1822, and it did so. Since then it has been seen regularly every 3.3 years, and the reappearance during 1964 was the forty-seventh observed return to perihelion.

Fittingly, the comet was named in Encke's honor, but it is more general nowadays to name a comet after its actual discoverer. Thus when Robert Burnham found a new comet in December 1959, it was officially known as Burnham's Comet. (It was, incidentally, visible without a telescope, and when at its closest to

MRKOS' COMET OF 1957, *photographed by E. A. Whitaker. This was the second reasonably bright comet of 1957; it was quite conspicuous to the naked eye for a few days in the autumn. It was discovered by the Czech observer Antonín Mrkos; the first observations from Britain were secured independently by a fifteen-year-old amateur, Clive Hare*

ALCOCK'S FIRST COMET OF 1959. *Drawing by G. E. D. Alcock*

the Earth, in late April 1960, was in the region of the north celestial pole.)

There are also special systems for numbering comets. Each time a discovery is made, it is given a year and a letter; the first new comet of 1959 was 1959a, the second 1959b, and so on. Burnham's, the eleventh and last, was 1959k. A year followed by a Roman numeral indicates that the comet has been designated according to its time of perihelion; the first 1960 comet to reach perihelion was 1960 I, and the next 1960 II.

Encke's was the first short-period comet, but nowadays nearly one hundred are known. All are faint, but their movements are so well known that astronomers always know when and where to expect them. During 1960, for instance, nine periodical comets, one of which was Encke's, were under observation at various times.

Comets have very small mass, and are strongly influenced by the planets, particularly Jupiter. Other planets, also, produce marked effects, but Jupiter is so much the most massive that it is always much the most important factor. The orbits may be highly inclined, and some comets, notably Halley's, move in a wrong-way or retrograde direction. Comets have been aptly termed 'wanderers in space'.

Amateur comet-hunters are still at work. It takes many hours of patient, systematic searching to make a discovery, and the unlucky enthusiast may continue unrewarded for years; a telescope with a good light-grasp and wide field of view is needed. Yet there is always the chance of spectacular success. In 1959 a British amateur G. E. D. Alcock, by profession a teacher in Peterborough, found two new comets within a few days. He has since found two more (to say nothing of three novæ).

Mention must also be made of E. E. Barnard, who found many comets during the last century, and the American amateur Leslie Peltier, who has so far made seven discoveries.

Great comets were seen fairly often during the period from 1700 to 1910. Donati's Comet of 1858 had a lovely, curved tail shaped like a scimitar; Chéseaux' Comet of 1744 had a multiple tail, and the comets of 1811 and 1843 stretched half-way across the sky. Halley's Comet is hardly 'great', but it and the Daylight Comet,

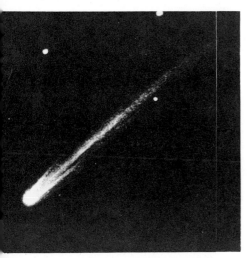

ALCOCK'S SECOND COMET OF 1959. *Drawing by G. E. D. Alcock*

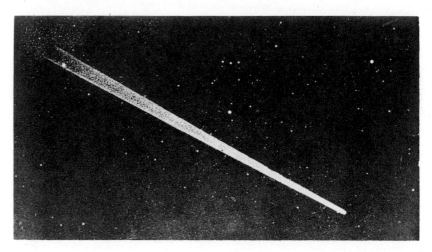

THE GREAT COMET OF 1811. *This woodcut shows one of the most spectacular comets ever observed*

BENNETT'S COMET, *photographed by its discoverer, the South African amateur astronomer, J. Bennett*

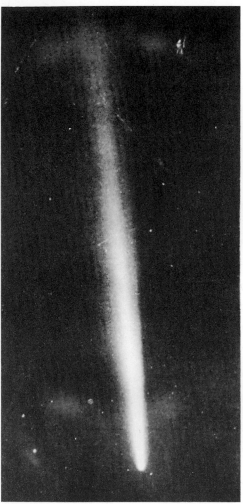

COMET IKEYA-SEKI, 1965, OCTOBER, *photographed by D. Andrews at the Boyden Observatory in Africa*

which also appeared in 1910, have so far been the last of their kind. The comet of autumn 1965, discovered by the Japanese amateurs Ikeya and Seki, became spectacular for a brief period as seen from some parts of the world, though from Britain it was a disappointment. However, Bennett's Comet of 1970, discovered by the South African amateur astronomer of that name, was much more imposing. Observers in the southern hemisphere saw it at its best, but even when it came north, and so could be seen from Europe, it was still a striking object even though it hardly merited the title of 'great'.

We cannot tell when the next really brilliant comet will appear; it may be at any time. Of all the naked-eye comets, only Halley's has a period of less than a century, so that it alone can be predicted.

It is natural to suppose that a comet moves head-first with its tail streaming out behind it, but this is by no means always true. When moving towards the Sun the comet goes head-first, but after it has passed perihelion and has begun its return journey the tail takes the lead.

The old explanation of this was that the tail was driven away from the comet by 'light pressure'. Light does indeed exert a force, though by everyday standards the pressure is extremely small; it was demonstrated in 1900 by the Russian physicist Lebedev, and his countryman Theodor Alexandrovitch Brédikhine supposed that the behavior of comet-tails could be accounted for in such a way—after all, the Sun is extremely brilliant, and the particles in a comet's tail, less than 1/100,000 of an inch in diameter, might be expected to be of just the size for radiation pressure to overcome the ordinary effects of gravity so far as the comet itself was concerned.

Brédikhine's theory, developed by Svante Arrhenius of Sweden and Karl Schwarzschild of Germany, seemed plausible enough, but recently it has been found to be inadequate. The true explanation seems to be more complex, involving magnetic effects and also the 'solar wind', a constant stream of particles emitted by the Sun. At any rate, a comet's tail points more or less away from the Sun, as shown in the diagram.

CHÉSEAUX' COMET OF 1744. *This was not one of the most brilliant of comets, but it was extremely spectacular, and was unique in developing a seven-fold tail. Ordinarily a great comet does not have more than one or two major tails, and nothing similar to Chéseaux' Comet has been seen since. The drawing given here is taken from an old sketch made at the time*

DRAWING OF BIELA'S COMET, *made in 1845. At this time the comet had separated into two distinct parts—the first time that such a phenomenon had been observed. In view of later events, it seems that this marked the beginning of the complete disintegration of the comet*

Nowadays, most astronomers support a theory due to F. L. Whipple, according to which a comet's head is made up mainly of 'ices'. When the comet nears the Sun, there is great evaporation, giving off gases which produce the *coma* around the comet's central part or *nucleus*. It has been said that a comet is mainly a collection of dirty ices, which is unromantic but probably true.

If a comet experiences evaporation each time it comes close to the Sun, a short-period comet, which returns every few years, cannot last for long on the astronomer's time-scale. This has been found to be the case. Several short-period comets seen at various returns during the last century have disappeared, though others (such as Holmes' Comet) were 'lost' for decades and then rediscovered. However, the most startling story concerns Biela's Comet.

In 1826 a new comet was found at almost the same time by Wilhelm von Biela, an officer in the Austrian Army, and Jean Gambart, assistant at Marseilles Observatory. It proved to have a period of $6\frac{3}{4}$ years. It was seen in 1832; missed in 1838, because it was badly placed for observation; and picked up again once more in 1845. Then it created an astronomical sensation by dividing into two parts, so that the appearance was, as G. de Vaucouleurs has put it, one of 'twin comets sailing through space in convoy'. In 1852 the twins were seen once more, rather more widely separated. In 1858 the conditions were again poor, but in 1866 they should have been good, and observers waited eagerly to see what would happen.

The answer was—nothing. The calculations were right, but in spite of the most intensive searches the comet did not appear. Something had happened to it.

The next return was expected in 1872. Again the comet was absent, but this time a spectacular meteor shower was seen in its place. Clearly the comet was dead, and the meteors represented the fragments.

Actually the full explanation is not so simple as this, but there is certainly a close link between comets and meteors. The fate of Biela's Comet proved it, and for years afterwards meteors were recorded whenever the comet should have been seen. Even today we still see a few 'Bieliid' meteors, though the shower is no longer brilliant.

It seems that as a comet moves round the Sun it leaves a trail of

BEHAVIOR OF A COMET'S TAIL. *When a comet is remote from the Sun its tail does not develop. As the comet nears perihelion, the tail may become very long, and it points more or less away from the Sun. When the comet has passed perihelion, it recedes from the Sun tail-first. A comet's tail always points approximately away from the Sun.*

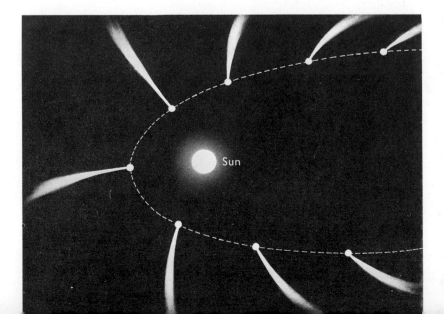

débris behind it, and this débris is gradually spread out all round the orbit. When the Earth passes through, the result is a meteor display.

We know that a shooting-star is simply a tiny piece of material which enters the Earth's upper air, moving at anything up to 45 miles per second, and is destroyed by friction as it speeds through the atmosphere. The 'shower' effect is simply one of perspective—just as in the case of parallel roads, which seem to meet at a point in the distance. The meteors are moving in parallel paths, and so seem to issue from a definite point or *radiant*.

More than a dozen fairly rich meteor showers are seen each year. Pride of place must go to the Perseids, which are so called because the radiant lies in the constellation Perseus. They occur

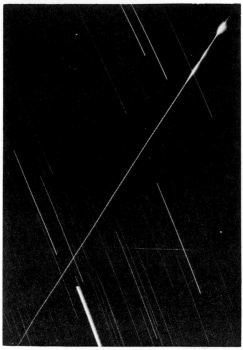

PERSEID METEOR, *photographed by H. B. Ridley on August 15, 1958*

PHOTOGRAPH OF THE MOTORWAY M1, *some miles outside London, taken from one of the bridges. It shows the parallel lanes apparently meeting at a point near the horizon. Photograph by Patrick Moore, 1960*

from about July 27 to August 17, and if the Moon is out of the way and the sky is dark they are often spectacular. Among other showers are the Quadrantids (January 3), the Orionids (October 20–21), and the Geminids (December 12–13).

Unusual displays are seen occasionally. A.D. 902 was named 'the Year of the Stars', because on one night it is said that the meteors appeared to be falling as thickly as snowflakes. The showers of 1366, 1799, 1833 and 1866 were almost equally startling. These latter showers were due to the Leonids, associated with a faint comet first seen in 1866 and named after its discoverer, the German astronomer Ernst Tempel.

The comet still exists, though it is very faint. The Leonid shower also exists, but we seldom see the best of it. Every 33½ years the Earth used to pass right through the swarm, but between 1866 and 1899 the shower was perturbed by the gravitational effects of the giant planets, and the orbit was so greatly altered that we no longer meet it. The expected showers of 1899 and 1933 failed to appear, but the Leonids are still very much in evidence, and there was a magnificent display in November 1966 as seen from parts of the United States.

Most meteors are smaller than a pin's head, and so are destroyed in the upper air. It is estimated that about 100 million of them enter

LEONID METEORS, *photographed from Arizona during the great shower of November 17, 1966. Photograph by D. R. McLean* the atmosphere each day. If you go outside on a dark, clear night when a shower is due, and spend a few minutes staring at the sky, you will be very unlucky if you do not see at least one shooting-star.

On April 26, 1803, at one o'clock in the afternoon, the inhabitants of the little French village of L'Aigle were disturbed by a strange sound. It was not unlike a violent roll of thunder—yet the skies were cloudless, and there was no sign of a storm. As the villagers rushed out of their houses in alarm they caught sight of an immense 'ball of fire' racing across the sky, and as it vanished there came a series of explosions which could be heard for 50 miles around. A few minutes later a great number of stones fell to the earth, landing at speeds great enough to cause them to bury themselves deeply in the ground.

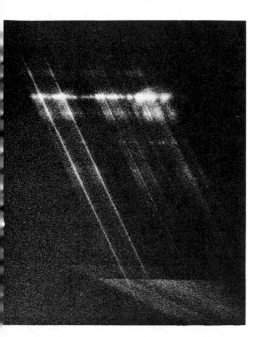

SPECTRUM OF A TAURID METEOR, *photographed by H. B. Ridley on October 29, 1954. This was the first meteor spectrum obtained in the United Kingdom*

EXPLODING ANDROMEDID METEOR, *photographed by H. Butler in 1895* ▶

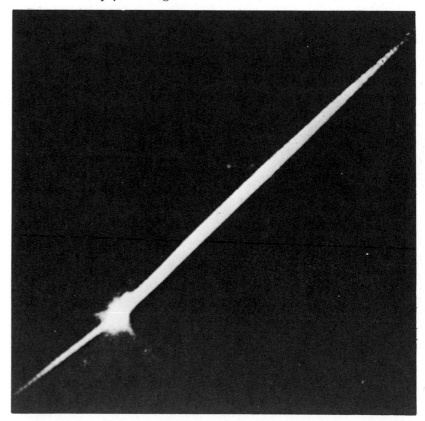

THE HOBA WEST METEORITE IN AFRICA. *The weight exceeds 60 tons*

The affair reached the ears of scientists in Paris, and a leading French astronomer, Jean-Baptiste Biot, was sent to investigate. He examined the strange stones, the heaviest of which weighed $17\frac{1}{2}$ pounds, and came to the conclusion that they had indeed come from space. This was the first time that astronomers had admitted the existence of *meteorites*; but once the discovery had been made, other meteorites were identified. One, for instance, was the 276-pound stone which had fallen at Ensisheim, in Alsace, in 1492. The famous sacred stone in the Holy City of Mecca also seems to be a meteorite.

Meteorites are relatively large bodies, ranging in size from pebbles to great blocks weighing tons. The largest ever seen to fall, over Arkansas in 1930, weighs 820 pounds, but a meteorite which

PIECE OF BARWELL METEORITE, 1965. *Photograph by Patrick Moore, who discovered this fragment in a field near Barwell village*

THE ZODIACAL LIGHT, *photographed at Chocaltaya, Bolivia, at an altitude of 17,000 feet, with the Sun 18 degrees below the horizon. The photograph was taken by D. E. Blackwell and M. F. Ingham. A single lens camera was used (aperture f/18, focal length 5 inches) with an exposure of 10 minutes, which accounts for the star trails on the plate*

fell at Hoba West in South Africa, presumably in prehistoric times, weighs 60 *tons*.

Such falls are extremely rare, and there is no proved case of anyone being killed by a meteorite, though admittedly there have been one or two narrow escapes. Meteorites do not come from shooting-star showers; they may indeed be more closely related to the asteroids. They are of two main types; aerolites (stony in composition), and siderites (largely nickel-iron). Most museums have collections of them.

A large meteorite can, of course, produce a crater. Such is the Meteor Crater at Coon Butte in Arizona, almost a mile across and 600 feet deep; others are the New Quebec crater (possibly meteoric), and smaller formations at various sites in Arabia, the U.S.A., Australia and the Baltic island of Oësel.

During the present century there have been only two major falls. One was that of June 30, 1908, when a large meteorite fell in Siberia and blew trees flat for twenty miles round the point of impact. The other occurred on February 12, 1947, also in Siberia. Fortunately the objects fell in uninhabited wasteland. If either had hit a big city the death-roll would have been colossal.

On Christmas Eve, 1965, an interesting stony meteorite fell at Barwell, in Leicestershire (England). It produced a brilliant 'fireball' as it came down through the atmosphere, and was seen over a wide area. Many fragments of it have been recovered; the piece shown here weighs over 1 lb.—I found it in a field when I visited the site later. The total weight of the original meteorite must have been about 200 lb., which is a record for Britain. Since then we have had the Bovedy Meteorite, which shot over England and Wales and deposited fragments over parts of Northern Ireland, though it seems that the main mass fell in the sea.

Lastly in our brief survey of the Sun's kingdom we come to the Zodiacal Light, a faint luminous band extending along the ecliptic. It is best seen after dusk in spring and before dawn in autumn, since the ecliptic then makes its steepest angle with the horizon. In Britain it is never very conspicuous, but is sometimes prominent from countries which have clearer skies. It and a faint light in the opposite direction, the *Gegenschein* or Counterglow, are thought to be due to meteoric bodies forming a lens-shaped cloud round the Sun, in the plane of the Earth's orbit. The Zodiacal Light was named by Cassini in 1683, but it had been previously recorded by several astronomers—including Kepler, who believed it to be the outer atmosphere of the Sun. The more elusive Gegenschein was discovered in 1854 by the Danish astronomer Theodor Brorsen.

During the last few chapters we have taken the story of the Solar System through from Herschel's time up to the present. We have not yet finished; the new rocket experiments have added to our knowledge, and are bound to lead to great new developments within the next few years. But before dealing with space research, let us turn back to the story of the stars.

WOLF CREEK CRATER *in Australia, which is almost certainly meteoritic. Photograph by G. J. H. McCall*

21   how far are the stars?

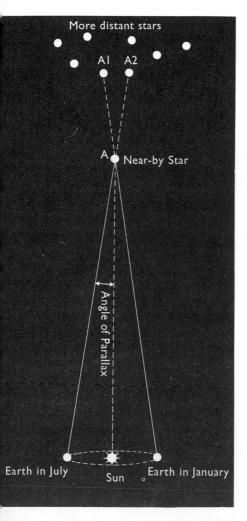

More distant stars

A1 A2

A Near-by Star

Angle of Parallax

Earth in July   Sun   Earth in January

THE PARALLAX METHOD OF MEASURING THE DISTANCE OF A STAR. *The relatively close star, A, seems to shift in position over the course of a year; in January it is seen at A1, in July at A2. However, it must be remembered that the diagram has had to be drawn completely out of scale. The parallax shifts are extremely small even for the closest stars, and it is not surprising that even Herschel failed to detect them. This method of trigonometrical parallax is practicable only for stars in the neighborhood of the Sun, since at greater distances of, say, a few hundred light-years, the shifts become so small that they are swamped by unavoidable errors in observation*

GIOVANNI CASSINI, the Italian astronomer who became first Director of the Paris Observatory, believed the Sun to be 86 million miles away. The actual distance is now known to be 7 million miles greater than this, but Cassini's estimate was certainly of the right order, and the work of Kepler and Newton made it possible to draw up a reasonably accurate map of the Solar System. Saturn, the outermost of the planets known in ancient times, proved to move round the Sun at a distance of nearly 900 million miles.

Mapping the Solar System had been difficult enough, but finding out the distances of the stars was much harder still. All that Cassini could say was that even the nearest star (excluding the Sun, of course) must be tremendously remote.

The first man to tackle the problem really seriously was the third Astronomer Royal, the Rev. James Bradley, who succeeded Halley in 1742 and held office until his death twenty years later.

Bradley, born in 1693, was educated at Oxford, and became Vicar of Bridstow in Monmouthshire. Then, in 1721, he went back to Oxford as Professor of Astronomy, and remained there until he moved to Greenwich when Halley died. Bradley decided to improve the existing star catalogues, and he did so very thoroughly, since his measures of the positions of over 60,000 stars were so good that even today they are of great value.

Yet Bradley's greatest discovery was made in 1728, before he became Astronomer Royal. The story is an interesting one, and shows how careful an observer Bradley was. It is connected, too, with his attempts to measure the distances of the stars by the method of *parallax*.

Suppose you hold a ruler at arm's length, and line it up with a distant object such as a tree, using one eye only. Now, without moving your head or the ruler, use your other eye. The ruler will no longer seem to be lined up with the tree; it will have apparently shifted, because it is being viewed from a slightly different direction. The angle by which it seems to have moved is a measure of parallax. If you know the distance between your eyes, and also the angle of parallax, it is possible to work out the distance of the ruler from the 'base-line', or line joining your eyes.

A similar method is used by surveyors who want to measure the distance of an object which cannot easily be reached. A longer base-line is needed, but the principle is exactly the same.

Bradley wanted to apply the parallax method to the stars. Suppose that we have one star, A in the diagram, which is relatively near? In January, we will be seeing it in position A1, compared with other more distant stars; six months later the Earth will have moved round to the other side of its orbit, 186 million miles away from its January position, and our star will appear in position A2. Measuring the angle of shift, and knowing the size of the Earth's orbit, will allow us to work out the distance of the star A.

STAR-CLOUDS IN THE MILKY WAY.
This photograph shows some of the richest regions of the Milky Way. Each point of light is a sun, perhaps far larger and more luminous than our own Sun. The distances of objects in these star-clouds are so great that the method of trigonometrical parallax cannot be used, and less direct (though probably no less reliable) means are employed

Using a special instrument owned by James Molyneux and erected in Kew, Bradley set out to measure the position of the star Gamma Draconis, in the region of the Great and Little Bears. He selected this particular star because it passes directly overhead in the latitude of London, and was thus convenient for measurement.

Bradley knew that the parallax shift would be very small, but his results puzzled him badly. There were shifts indeed, but they did not seem to be due to parallax; Gamma Draconis appeared to be moving in a tiny circle, returning to its original position after a period of one year. Bradley checked with other stars, and found that they all behaved in the same way.

According to a famous tale which may well be true, Bradley hit on the solution one day when he was out sailing on the Thames. He noticed that when the direction of the boat was altered, the vane on the mast-head shifted slightly, even though the wind remained the same as before. At once he realized that this was the principle which accounted for the behavior of Gamma Draconis. Light, as Rømer had found, does not travel instantaneously, but at a definite speed of 186,000 miles per second. The Earth also is in motion, travelling round the Sun in a more or less circular path at an average speed of some $18\frac{1}{2}$ miles per second, and so the Earth's 'direction' is changing all the time. Let the Earth represent the boat, and the incoming light represent the wind, and the position becomes clear; there is always an apparent displacement of a star towards the direction in which the Earth is moving at that moment. This is *aberration*.

Bradley had made an important discovery, but he had not succeeded in his original object. Herschel, many years later, was equally unsuccessful, but—like Bradley's—his work led to an unexpected discovery. He had selected double stars for measure-

POSITION OF GAMMA DRACONIS.
This was the star studied by Bradley in an attempt to determine its parallax. Gamma Draconis was selected because it passes directly overhead at the latitude of Greenwich, and was thus convenient for observation with the special instruments used by Bradley. The parallax was not detected, but Bradley's research led to his discovery of the aberration of light

ment; if one member of the pair were much closer than the other, then its parallax shift, relative to the more distant component, would be comparatively easy to measure. Nothing of the kind was found. Instead Herschel discovered that in some cases, at least, the components appeared to be in relative motion round their common centre of gravity.

John Michell, an English amateur who was born in Nottingham in 1725 and died in 1793, had suggested that many double stars might be genuinely associated, so forming true pairs or *binaries*. Herschel's measures showed that Michell had been right, but the star-distances remained as much of a problem as ever.

After Herschel's death in 1822 the matter was taken up by three astronomers—Friedrich Bessel, formerly Schröter's assistant and by now Director of the Königsberg Observatory in Germany; F. G. W. Struve, using the famous Fraunhofer refractor at Dorpat; and Thomas Henderson, at the Cape of Good Hope. They worked quite independently of each other, and selected different stars.

Bessel decided upon a 5th-magnitude star in Cygnus (the Swan), numbered 61 in Flamsteed's catalogue and therefore known as 61 Cygni. It is a binary, with the two components far enough apart to be separated with a small telescope, but Bessel was interested in it mainly because of its exceptionally large *proper motion*. Relative to the other stars, it moves over 5 seconds of arc per year. This means that it takes 350 years to shift by an amount equal to the apparent diameter of the Moon; even so, it is unusually fast-moving, and Bessel decided (rightly) that it must be one of our nearest stellar neighbors.

He began work in 1837, and only a year later he was able to announce that both components of 61 Cygni showed a parallax of 0·3 seconds of arc, giving a distance which amounted to roughly 60 million million miles. His measures were excellent, and the values which he gave are almost the same as those accepted to-day. Yet the parallax shift of 0·3 seconds per arc is much less than the apparent diameter of a coat-button seen from a distance of ten miles.

Faced with distances of this kind, the mile becomes inconveniently short as a unit of length; it is rather like giving the length of a journey between London and New York in centimetres. Fortunately a much better unit is to hand. Light moves at 186,000 miles per second, and in one year it will therefore cover 5,880,000,000,000 or nearly 6 million million miles. This is the astronomical *light-year*. Bessel measured the distance of 61 Cygni at just under 11 light-years, which is very close to the modern value.

Strictly speaking, Bessel's measures were made after those of Henderson, who had been studying the bright southern star Alpha Centauri. Henderson, born in Dundee in 1798, was appointed H.M. Astronomer at the Cape in 1832, and his work on parallax had been carried out during his thirteen months' stay there. He had had to resign and return home because of ill-health, and he did not finish the calculations until after Bessel's announcement. Alpha Centauri proved to be closer than 61 Cygni, since it lay only 4⅓ light-years off.

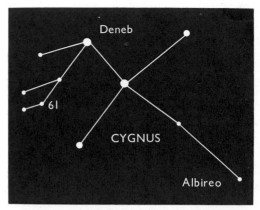

POSITION OF 61 CYGNI. *The star numbered 61 in Cygnus by Flamsteed was the first object outside the Solar System to have its distance measured; the work was carried out by Bessel, and published in 1838. 61 Cygni is of the 5th magnitude, and is therefore visible to the naked eye, but it is far from conspicuous.. It is a relatively wide binary, and has an exceptionally large individual or proper motion, which was why Bessel concentrated his attention on it*

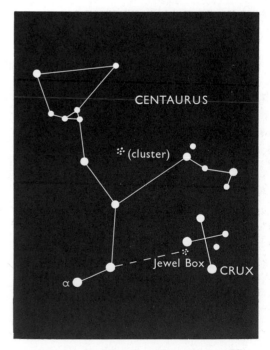

POSITION OF ALPHA CENTAURI. *The bright southern star Alpha Centauri was studied for parallax shift by Henderson, at the Cape of Good Hope. Measurable shifts were found, and Henderson was able to give a reliable figure for the star's distance, which amounts to rather more than 4 light-years. Alpha Centauri is thus the nearest of the brilliant stars. It is a splendid binary, and a fainter member of its system, Proxima, is the nearest known star to the Sun.*

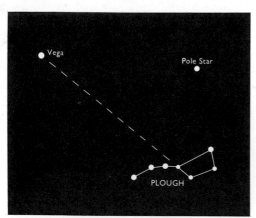

POSITION OF VEGA. *O. Struve made efforts to measure the distance of Vega at about the same time as Bessel and Henderson were carrying out their researches. Struve's result was less accurate, mainly because Vega is relatively remote. It is the fifth brightest star in the sky, and is easy to find, both because of its brilliancy and because of its decidedly bluish color; the Great Bear may be used as a direction-finder to it. Vega is the chief star of the small but interesting constellation of Lyra, the Lyre. It is fifty times as luminous as the Sun*

THE SCORPION. *The leading star of Scorpio, the Scorpion, is Antares, known as 'the rival of Mars' because of its strong reddish color. Antares is a typical Red Giant star, of relatively low surface temperature but immense size. Its distance from us is about 360 light-years, and it is accompanied by a faint companion which is decidedly greenish. Scorpio is a brilliant constellation, but is never seen to advantage in Great Britain or the northern United States because of its southern position; from London and New York, the 'tail' never rises at all*

Struve had had the hardest task, since his selected object was the lovely bluish star Vega, in Lyra (the Lyre), which may be seen almost overhead in London or New York during summer evenings. Vega is 26 light-years away, so that the parallax is smaller, and it is not surprising that Struve's results were less accurate.

Alpha Centauri is a glorious binary. Associated with it is a third, much fainter star, known as Proxima because it has proved to be the closest known—4·3 light-years away. Altogether nineteen stars are known to lie within a dozen light-years of us, including Sirius (8·6 light-years) and Procyon (11 light-years).

To show how vast these distances are, it will be helpful to describe a scale model. The 93 million miles between the Earth and the Sun is known as the *astronomical unit*, and there are 63,310 astronomical units in a light-year. It so happens that there are 63,360 inches in a mile, so that the ratio is very much the same. Let us take a scale in which the Earth–Sun distance is given as one inch. On this scale, Proxima Centauri will be over 4 miles away: Sirius 8·6 miles, 61 Cygni about 11 miles, Vega 26 miles and so on. Yet Pluto, the outermost planet in the Sun's family, will be only about 3 feet away. The Solar System is indeed a small place on the astronomer's scale.

Nowadays the distances of many stars have been measured, and the results are staggering. Capella in Auriga (the Charioteer), which appears almost as bright as Vega, is 47 light-years away; Antares in Scorpio (the Scorpion) 360 light-years; Rigel in Orion perhaps as much as 900 light-years, and so on.

It is interesting to find, then, that as seen from Earth Rigel appears almost of the same brightness as Vega and Capella. Since it is much more remote, it must be much more luminous. We now know that it shines with at least 50,000 times the candle-power of the Sun; even so, we know of many stars which are more luminous still.

Unfortunately the parallax method can be used for only the relatively near stars. By 150 light-years, the shifts have become so small that they are hard to measure properly, and by 600 light-years they are swamped by unavoidable errors in observation, so that the whole system breaks down. For more remote objects, astronomers have to turn to less direct methods, which means using instruments based on the principle of the spectroscope.

Here we must begin another story—that of the analysis of starlight; and as so often happens, we find that it starts with the work of Isaac Newton during the years when he was at Woolsthorpe, far from Plague-stricken London. But it is easy to see that yet another astronomical landmark was passed in 1838, when the patient work of Bessel, Henderson and Struve first proved the vastness of the universe.

THE CONSTELLATION OF ORION, *photographed by Henry Brinton*

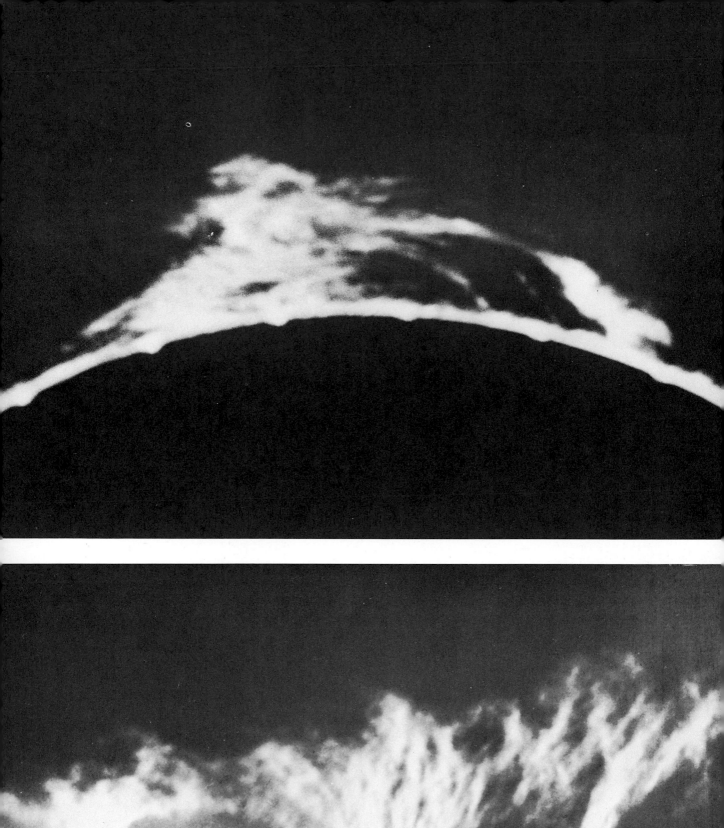

22   exploring the spectrum

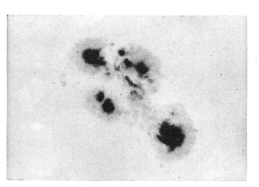

SUNSPOT GROUP, *photographed on March 31, 1960 by W. M. Baxter, using a 4-inch refractor*

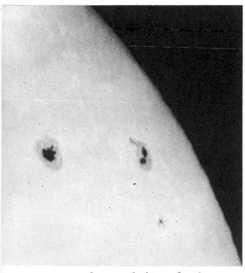

SUNSPOTS, *photographed on October 17, 1959 by W. M. Baxter, using a 4-inch refractor. Note the foreshortening of the group nearer the Sun's limb*

SOLAR PROMINENCES. (Upper) *Prominence 132,000 miles high, photographed in calcium light on August 18, 1947. (Lower) Prominence 80,000 miles high, photographed in calcium light on August 21, 1909. Photographs by Mount Wilson and Palomar Observatories*

IN 1666 ISAAC NEWTON, then a young and unknown Cambridge student, used a prism to produce a rainbow *spectrum* from a beam of sunlight. This was the first indication that what we call 'white' light is really a mixture of colors, and it proved to be one of the most important discoveries in astronomical history.

In 1802 William Hyde Wollaston, another Englishman, noticed seven dark lines in the solar spectrum. Wollaston, a native of East Dereham in Norfolk, had taken a medical degree at Cambridge, and had become known as a man with a wide knowledge of all branches of science. Yet on this occasion he missed a great opportunity. He thought that the dark spectral lines merely indicated the boundaries between various colors, and paid no more attention to them.

Fraunhofer, the penniless orphan who rose to become the most skilful instrument-maker of his day, saw the dark spectral lines once more about 1814. Unlike Wollaston, he inquired further, and found that the lines never seemed to change; whenever he examined the Sun's spectrum, the familiar dark markings appeared, always in exactly the same positions. Fraunhofer measured the positions of over 500 of them, and even today they are known as 'Fraunhofer Lines' in his honor.

The lines showed some particularly interesting features. For instance, the solar spectrum revealed a prominent dark double line in the yellow section. Luminous sodium vapor, however, showed a *bright* double yellow line, and Fraunhofer wondered whether there might be some connection. Given enough time he might well have hit upon the solution, and it is tragic that he died while still a young man.

The mysterious dark lines were accounted for in 1859 by Gustav Robert Kirchhoff, who had been born at Königsberg in 1824, had been educated at Berlin and Marburg, and had been appointed Professor of Physics at Heidelberg University. To show Kirchhoff's reasoning, it will be necessary for us to say something about the way in which matter is built up.

All matter is composed of *atoms*, which combine into groups or *molecules*. Atoms are far too tiny to be seen individually by ordinary means, and they are of different types. First let us consider the simplest of all, the atom of hydrogen.

Hydrogen is the lightest substance known, and is also the most plentiful substance in the universe. Its atom may be said to be made up of a central particle or *nucleus*, around which revolves another particle or *electron*. The nucleus itself consists of a *proton*, which carries a unit positive charge of electricity. To balance this the circling or 'planetary' electron carries a unit negative charge; this cancels out the effect of the proton, and so the complete atom is electrically neutral.

The next lightest substance is another gas, helium. This time the nucleus is rather more complex, and carries two positive

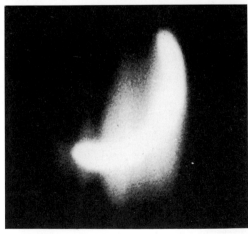

THREE STAGES OF THE SOLAR ECLIPSE OF FEBRUARY 15, 1961. *These photographs were taken from an R.A.F. Shackleton aircraft flying at 10,500 feet over the Bay of Biscay. The first two pictures were taken before totality, and the third one after totality. Conditions were not good, and some high-altitude cloud shows up on the photographs*

charges. There are however two planetary electrons, and once again the complete atom is electrically neutral.

Lithium has three planetary electrons; consequently there are three charges in the nucleus—and so on. Each substance, or *element*, has an extra electron. Oxygen, for instance, has eight. As the number of electrons goes up, the atom becomes more and more complex, until finally we come to uranium, with ninety-two electrons.

All the matter known to us is made up of these ninety-two fundamental elements, and we may be certain that no more remain to be discovered. It is impossible to have half an electron, and so there is no room in the sequence for any new elements. It is true that artificial elements have been made, with more than ninety-two electrons, but these are unstable and probably do not occur in nature.

It may seem surprising that there are so few elements, but of course they combine in many different ways. Water, for instance, is made up of hydrogen and oxygen. Two atoms of hydrogen combine with one of oxygen to make up a water molecule, giving the chemical formula H_2O. Salt is composed of one sodium atom together with one of chlorine, and so on. There is some sort of analogy here with writing; all the thousands upon thousands of words in the English language are composed of only twenty-six fundamental letters of the alphabet, from A to Z.

Kirchhoff announced three Laws which still bear his name. These laws form the basis of all spectroscopy, astronomical or otherwise, and give a complete explanation of the Fraunhofer lines.

The first law is easy enough. It states that incandescent solids, or incandescent gases under high pressure, produce a rainbow band or *continuous* spectrum. There is a full range, from red at the long-wave end of the band down to violet at the short-wave end.

The second law states that a luminous gas or vapor under low pressure will produce not a rainbow band, but merely a number of isolated bright lines. This is an *emission* spectrum. The vital fact is that each element will produce its own distinctive set of lines. The double yellow line of sodium, seen by Fraunhofer, cannot possibly be produced by any element except sodium; lines due to hydrogen cannot be due to anything except hydrogen, and so on. Each element has its own particular trade-marks, which cannot be duplicated. Often the spectrum is very complex, and iron alone produces many hundreds of lines, but careful measurement will usually enable the research worker to disentangle one line from another.

The crux of the whole dark-line problem is Kirchhoff's third law. The best way to explain it is to describe a simple experiment. If you burn salt in a flame, you will produce an emission spectrum, including the double yellow line. This line is due to sodium; salt, as we have seen, is made up of sodium and chlorine. If you look at the spectrum of an electric light bulb, you will see a rainbow, since the filament of the bulb is an incandescent solid—which, following Law 1, must give a continuous spectrum. Now take the bulb and put it behind the flame, so that you are examining the flame against the background produced by the bulb. Instead of a

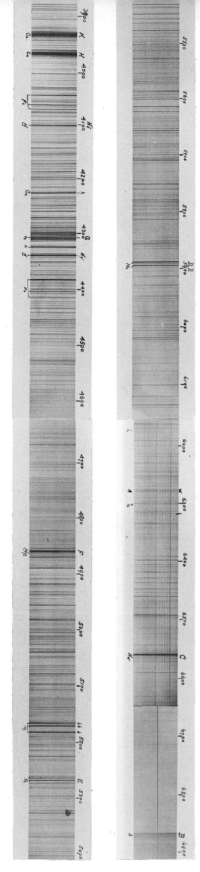

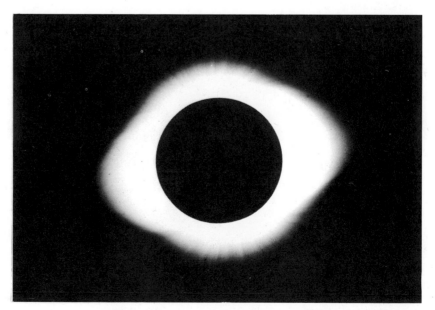

rainbow crossed by bright lines, what you will see takes the form of a rainbow crossed by *dark* lines. The atoms in the sodium vapor are removing part of the corresponding portion of the continuous spectrum, producing dark *absorption lines*. As soon as the background bulb is removed, the absorption lines due to the sodium vapor flash out and become brilliant once more. Yet their positions in the spectrum are unchanged.

In the Sun, the luminous surface or photosphere takes the place of the bulb; since it is incandescent high-pressure gas, it produces a rainbow. In front we have the luminous gas above the Sun, known as the chromosphere. By itself, the chromosphere would yield emission lines; but since the rainbow is behind it, these lines are reversed, and appear dark. They are in fact the lines seen by Fraunhofer. Among them is the celebrated yellow double. Since this must be due to sodium, we can prove that there is sodium in the Sun.

By now, some seventy of the ninety-two elements have been identified in the solar spectrum. One of them was found in the Sun before it was known on Earth. In 1869, an English astronomer named Norman (afterwards Sir Norman) Lockyer found a Fraunhofer line in the orange-yellow region which he could not identify. He suggested that it might be due to an unknown element, and proposed to name it helium, after the Greek word for 'sun'. A quarter of a century later, another Briton, W. R. Ramsay, discovered helium on Earth.

Lockyer had already made one important discovery, since in the previous year he had been one of the first men to observe solar prominences without waiting for an eclipse. Actual priority must go to a French astronomer, Pierre Jules Janssen, but the methods of the two astronomers were much the same.

In 1868 there was a total solar eclipse. As soon as the Moon hid the Sun completely, there was the usual appearance of the prominences, which looked like 'red flames' rising from the Sun's photosphere. These prominences had been seen at many previous

PART OF THE SOLAR SPECTRUM, *taken with the 13-foot spectrograph at Palomar. Many lines are shown, and the complexity of the spectrum is very evident. High dispersion or 'spreading-out' is possible, since in the case of the Sun there is plenty of light available*

TOTAL ECLIPSE OF THE SUN, JUNE 30, 1954, *photographed from Sweden by Professor Åke Wallenquist. This eclipse was just total from the northernmost Scottish islands, but was seen over most of Britain as a large partial. The line of totality extended across Norway and Sweden*

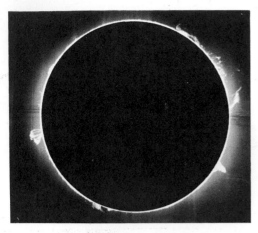

THE SOLAR CORONA, *photographed by the English Astronomer J. Jackson, from Giggleswick, at the total eclipse of June 29, 1927*

THE SOLAR FLARE OF AUGUST 8, 1937, *as photographed at Mount Wilson. (Left) Ordinary photograph of sunspot group. (Middle) the same region taken with the red hydrogen line H-alpha, showing the flare. (Right) five minutes later, again in H-alpha light, when the flare was at its maximum. Flares are now known to be common phenomena, but very few are visible in ordinary light*

total eclipses, but at all other times they are hidden by the general glare. By using spectroscopes to isolate the light due to the prominences, Janssen and Lockyer made it possible to study them under ordinary conditions. Years later, in 1891, George Ellery Hale, architect of the great reflectors, designed the *spectroheliograph*, an instrument for photographing the Sun in the light of one element only. Nowadays, monochromatic filters have come into general use.

Prominences are of two main types. Quiescent prominences may persist for many days, and sometimes measure hundreds of thousands of miles along their base, rising to as much as 30,000 miles above the photosphere. Eruptive prominences may move at speeds in the region of 250 miles per second, and have been seen at heights of half a million miles from the Sun's surface. They move so quickly, indeed, that moving pictures have been taken of them by French and American astronomers.

We now know a great deal about the way in which prominences behave, since there is no difficulty in keeping them under observation. So long as they could be seen only during eclipses, however, knowledge was bound to be limited. A total solar eclipse is a comparatively rare event, and no eclipse can last for more than a few minutes.

Beyond the chromosphere, where the Fraunhofer lines are produced, comes the glorious pearly corona. The great French astronomer Bernard Lyot, who was born in 1897 and died suddenly and tragically in 1952 when on an eclipse expedition to Africa, devised a special instrument for studying the corona without waiting for the Moon to hide the Sun; but though the innermost regions have been seen by such methods, we have to admit that our best opportunities are still confined to the fleeting moments of a total eclipse.

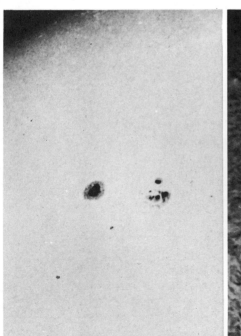
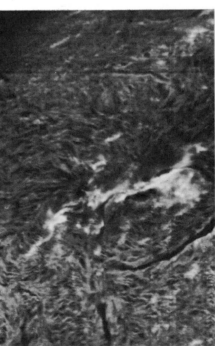
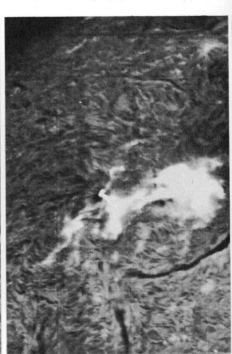

Telescopes can show sunspots and give us a good view of the solar surface, but by themselves they can tell us very little about the way in which the Sun is made up. Combined with spectroscopes, they can give us an excellent picture of what the Sun is really like. It is necessary to examine the spectrum in as much detail as possible; and since there is plenty of light available, special 'tower telescopes' have been built, notably at Mount Wilson and at the Italian Observatory of Arcetri, near Florence. With these instruments, the sunlight is concentrated by an upper mirror, and a spectrum produced in a convenient observing position.

It would take many pages to give anything like a proper description of the various instruments based on the spectroscope. For instance, the splitting-up of the light is often done not by a prism, but by a very finely-ruled *grating*. This was yet another development due to Fraunhofer, but such gratings are not easy to make, since they have to be ruled to many hundreds of lines per inch. Meanwhile let us turn back to stars, where the problems are much the same but the methods of study have to be altered somewhat.

The Sun, as we know, is an ordinary star. If it produces an absorption spectrum with a continuous background crossed by Fraunhofer lines, other stars may be expected to give similar effects. One of the pioneers in this field was Sir William Huggins, who was born in London in 1824, and established his own observatory in Tulse Hill. Helped by his wife, who was also a skilful astronomer, he examined the spectra of many stars, and in 1863 he identified various elements in Betelgeux and Aldebaran.

Unfortunately matters are made more difficult by the fact that the stars are relatively faint. There is no question of using tower-type instruments to give spectra of high dispersion; there simply is not enough light. Huggins made another step forward in 1863 when he first photographed the spectrum of a star, but even so he was facing grave problems. In those days photography was still rather primitive.

At about the same time, Angelo Secchi in Italy was carrying out work in connection with stellar spectra. While Huggins concentrated on a few stars, and studied them in as much detail as he could, Secchi did his best to examine large numbers of stars and fit them into different spectral types. He realized that all stars are not the same as the Sun; some are hotter, others cooler. This is shown at once by the obvious differences in color.

To the casual observer all stars may seem white, but this is far from being the case. Compare the two leaders of Orion, for instance; Betelgeux is clearly reddish; while Rigel, in the Hunter's foot, is white or even slightly bluish. Capella in Auriga is yellow, while Arcturus in Boötes (the Herdsman), is a glorious orange. White heat is greater than red heat, and therefore white or blue-white stars such as Rigel must have surface temperatures higher than those of the yellowish Capella or the orange Arcturus, while Betelgeux will be cooler still. Temperatures are bound to have great effects upon the type of spectrum produced.

Secchi divided the stars into four spectral classes. Type I was made up of white stars, II of yellow or orange, and III and IV of

THE TOTAL ECLIPSE OF THE SUN, JUNE 30, 1954, *as seen from Lysekil in Sweden. Conditions were fairly good, though there was a certain amount of very thin cloud. This was one of the most favorable European eclipses of recent years, comparable with the eclipse of 1961. The track of totality extended across Norway and Sweden, into Russia. However, expeditions sent to Scandinavia were in general handicapped by cloud, and from some sites no results were obtainable. It will be many years before another total eclipse will be seen from Scandinavia*

SOLAR PROMINENCE. *The whole edge of the Sun, photographed at Mount Wilson on December 9, 1929, in calcium light (K line)*

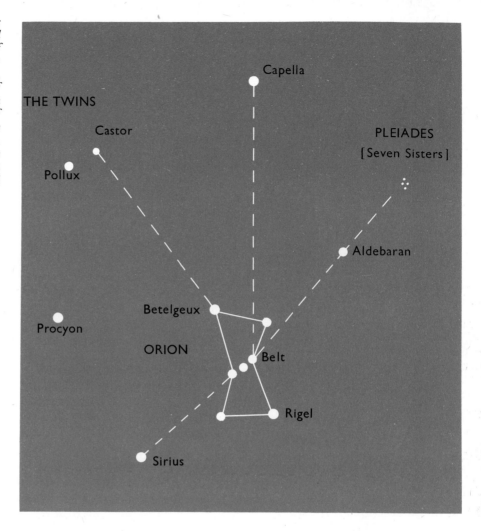

ORION AND HIS RETINUE. *Various types of stars are shown. Betelgeux is a Red Giant of type M; Aldebaran, also red, is of type K; Capella, type G, similar to the Sun; Procyon, slightly yellow, type F; Sirius, white, type A; Rigel, a particularly luminous star of type B. The remaining bright stars in Orion are also of type B. The two Twins are of different spectral type. Pollux is an orange K-star, while Castor is white and of type A. Some of the star-colors are noticeable with the naked eye, but are much better seen with optical aid; a pair of good binoculars will bring them out excellently*

THE 150-FOOT TOWER TELESCOPE AT MOUNT WILSON, *viewed from the north-east. This telescope is designed specially for solar work. Tower telescopes are, in fact, used exclusively for studying the Sun, and are able to produce spectra of very great dispersion. A similar instrument has been set up at Arcetri, in Italy—an observatory which has a great reputation in this field of research—and there are others in America and in Russia*

red. In 1890 E. C. Pickering at Harvard, drew up a more detailed classification in which he started off by dividing the stars into groups and lettering them A, B, C, D and so on, beginning with white stars and working through yellow, orange, and orange-red to red. As so often happens, the letters became out of order, and the final result was rather jumbled from the alphabetical point of view. However, Pickering's scheme is still used. The 'spectral alphabet' is: W, O, B, A, F, G, K, M, R, N, S.

Each type is again divided up into sub-classes, numbered from zero to 9. This gives a smooth sequence. To take part of the order at random, let us consider class A, so that we have A1, A2, A3 . . . A9. Type A1 is little different from B9, while A9 is almost the same as F0.

The spectrum of a star is not easy to study properly even with the help of photography, and to work out a reliable system took many years. Stars placed in class W are unusual because their spectra contain bright emission lines; they are known as Wolf-Rayet stars in honor of two astronomers who drew attention to them. Stars of types R, N and S are deep orange-red, and have relatively low surface temperatures of around 3,000 degrees;

all of them are very remote, and so appear faint in our skies. The best way to summarize the remaining classes, which make up the vast majority of the stars, is by a table:

TYPE COLOR		SURFACE TEMP. °C		TYPICAL STAR(S)
O	White	35,000	Both bright and dark lines.	Zeta Argûs (O5)
B	Bluish	25,000	Helium lines prominent.	Alkaid in the Great Bear (B3)
A	White	11,000	Hydrogen lines prominent.	Sirius (A1)
F	Yellowish	7,500	Calcium lines prominent.	Procyon (F5)
G	Yellow	6,000	Some lines due to metals.	Capella (G0); Sun (G2)
K	Orange	4,200	Stronger metallic lines.	Arcturus (K0)
M	Orange-red	3,000	Complex spectra, with bands due to molecules.	Betelgeux (M2)

This may seem straightforward enough, but in fact it is nothing of the kind. There are so many factors to be taken into account. For instance, stars of type B, such as Alkaid—the end star in the Great Bear's tail—show prominent helium lines, but this does not necessarily mean that they are very rich in helium; it may simply be that conditions are suitable for helium to show itself conspicuously. Then there are various stars which refuse to fit into any particular class, and have to be noted as 'peculiar'.

Yet it is possible to obtain a tremendous amount of information. For instance, studying a star's spectrum may often give a reliable clue as to how luminous the star really is. As soon as we know this, we can compare its apparent brightness with its true luminosity, and so work out its distance from us. Methods of this kind have to be used for all except our relatively near stellar neighbors, since parallax measures of the kind used by Bessel, Henderson and Struve are useless beyond a range of a few hundreds of light-years.

As recently as 1825 a French writer named Auguste Comte made the definite statement that mankind could never find out anything about the chemistry of the stars. Within half a century he had been proved wrong, but stellar spectroscopy led to all manner of other discoveries as well. In particular, Huggins, and (independently) the German astronomer Hermann Vogel, director of a private observatory at Bothkamp, made pioneer measures of *radial velocities*, using spectroscopic methods.

We know that the stars show slight but definite proper motions, so that over the centuries the constellation patterns will alter. Halley had been the first to detect these, when he found that three bright stars had moved appreciably since the days of Hipparchus. But what about the 'towards and away' movements? These would not reveal themselves as proper motions. A man walking straight towards you will not shift against his background, but will simply appear larger and larger as he draws near. Similarly, an approaching star will become steadily brighter, but the increase in brilliancy will not be enough to be measured except over a period of thousands of years. Huggins and Vogel made use of a principle which

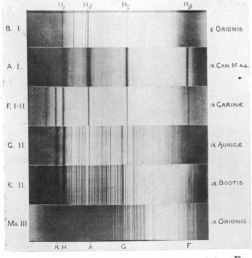

TYPICAL STELLAR SPECTRA, *types B to M. The spectra shown are of Alnilam (Epsilon Orionis, the middle star of Orion's Belt); Sirius (Alpha Canis Majoris), Canopus (Alpha Argûs, alternatively known as Alpha Carinæ); Altair (Alpha Aquilæ); Arcturus (Alpha Boötis) and Betelgeux (Alpha Orionis)*

had been laid down in 1842 by the Austrian physicist Christian Doppler and is still known as the Doppler Effect.

If you listen to a train which is approaching you, and sounding its whistle, the note will be high-pitched. More sound-waves are entering your ear per second than would be the case were the train standing still, and this raises the note. After the train has passed by, and has begun to recede, fewer sound-waves per second reach your ear, and the note drops. Light may be regarded as a wave-motion, and the effect here is to shift the spectral lines—to the violet or short-wave end if the source of light is approaching, to the red or long-wave end if it is receding.

Suppose we have a star which is coming towards us? More light-waves per second will reach us than would be the case for a star at rest, and all the spectral lines will be moved slightly towards the violet end of the rainbow. Similarly, a receding star will show a red shift. The amount of the displacement gives a key to the velocity. In this way the radial velocities of many stars have been measured. Sirius is approaching us at five miles per second, Vega at eight, Altair at sixteen; Capella is receding at eighteen miles per second, Betelgeux at thirteen, Aldebaran at thirty-four—and so on. By combining these radial velocities with the measured proper motions, as shown in the diagram, we can decide how the star is really moving through space with respect to the Sun.

Spectroscopy is also used for studies of the planets. It was by such methods that astronomers detected carbon dioxide in the atmosphere of Venus, and hydrogen compounds round the giant planets Jupiter, Saturn, Uranus and Neptune. Yet it is in exploring the stars that spectroscopy becomes our main research tool, and stellar astronomy has now turned into true *astrophysics*.

It is more than a hundred years since Kirchhoff laid down his three laws which opened the gateway to our knowledge, but the story began far earlier than that. It began, indeed, on the day when Isaac Newton passed sunlight through a glass prism and produced a colored rainbow.

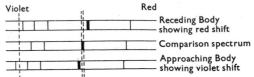

Violet   Red
Receding Body showing red shift
Comparison spectrum
Approaching Body showing violet shift

THE DOPPLER EFFECT. *In the upper diagram the light-source is assumed to be receding, and the spectral lines are shifted toward the red. The centre strip is for a stationary light-source, drawn in for comparison; the bottom strip, for an approaching source, shows a shift to the violet. This Doppler Effect is invaluable in astrophysics, since it enables the radial motions of celestial bodies to be measured. It must, of course, be borne in mind that the red and violet shifts are never easy to measure with precision, since there are many complications to be taken into account, but the results obtained are of a high degree of reliability. Doppler effects also enable the rotations of stars to be measured; as the star turns, the light coming from the approaching limb has a violet shift, while the light from the receding limb has a red shift. The result is that the line appears broadened, and the amount of the broadening gives a key as to the rate of rotation*

STELLAR SPECTRA. *When it is desired to study the spectra of many stars, as is necessary for many astrophysical investigations, it obviously saves a great deal of time to be able to photograph a number of spectra on a single exposure. On this photograph, taken by H. B. Ridley, each star is drawn out into a spectrum. Methods of this sort are extensively used in modern work, and have yielded very satisfactory results, though naturally they have their limitations*

23   giants and dwarfs of the sky

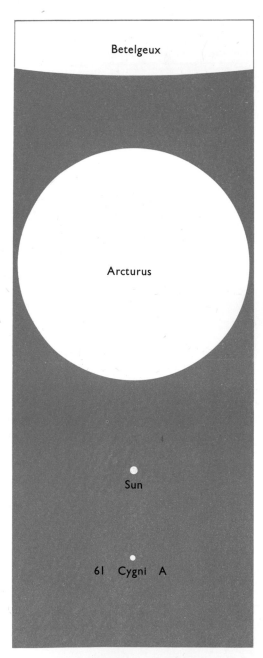

GIANT AND DWARF STARS, *drawn to scale. Betelgeux has a diameter of 250,000,000 miles; Arcturus 26,000,000 miles; the Sun 865,000 miles; 61 Cygni A 600,000 miles. The supergiant Betelgeux and the red dwarf 61 Cygni A each have M-type spectra. Arcturus is of type K, and the Sun type G*

BY THE EARLY YEARS of the twentieth century, astrophysics had become one of the most important branches of modern astronomy. The spectra of the stars had been studied and radial velocities measured according to the Doppler principle.

It was during this period that a very interesting discovery was made by a Danish astrophysicist, Ejnar Hertzsprung, who had been born near Copenhagen in 1873 and had tackled problems of celestial spectroscopy. Hertzsprung noticed that the red stars, most of which are of type M according to the Harvard classification, fell into distinct groups. There were very luminous red stars far brighter than the Sun, and very dim red stars much feebler than the Sun, but there did not seem to be any of 'middling' luminosity.

Much the same results were obtained independently by Henry Norris Russell, Director of the Princeton Observatory in America. To Russell we owe the terms 'giant' and 'dwarf' as applied to the stars. It became clear, too, that there was a giant and dwarf subdivision with the orange and the yellow stars, but not with the hotter white and blue stars of types B and A, most of which are extremely luminous.

Arcturus, the lovely orange leader of Boötes the Herdsman, is of type K. The brighter component of the 61 Cygni binary, famous as being the first star to have its distance measured by the parallax method, is also of type K. Yet here the resemblance ends. Arcturus is 26 million miles in diameter and 100 times as luminous as the Sun; 61 Cygni has a diameter of only 600,000 miles and only 6/100 of the Sun's luminosity. Arcturus is a true giant, while 61 Cygni is undoubtedly a dwarf on the cosmical scale. The mass ratio is only 16 to 1, so that the dwarf is the denser of the two.

If we turn next to stars of type M, we have a splendid example of a giant—Betelgeux in Orion, which is some 250 million miles across, larger than the Earth's orbit round the Sun. To show how vast Betelgeux is, it may be helpful to imagine a journey round it, moving at a steady 60 m.p.h. Anyone who had begun such a trip in the days of King Canute would still not have been once right round the star, since the time required for one 'lap' would be well over a thousand years.

Barnard's Star—named in honor of E. E. Barnard, who paid special attention to it—is one of our nearest stellar neighbors. Like Betelgeux it has an M-type spectrum, but its diameter is less than 150,000 miles. The difference in brilliancy is just as marked. Betelgeux shines with a luminosity equal to 1200 Suns; Barnard's Star has only 16/100 of the Sun's brightness.

What of the Sun itself? To some people it comes as a surprise to find that it is ranked as a Yellow Dwarf. True, it has a diameter of 865,000 miles; but Capella—a typical Yellow Giant—is more than 10 million miles across, and 150 times as luminous. However, it is among the white and bluish-white stars that we find the real searchlights of the universe. Rigel, as we have seen, is about 50,000 times as luminous as the Sun, and the most energetic stars known to us have something like a million Sun-power.

Hertzsprung and Russell worked away at the giant-and-dwarf problem, and the result was a diagram of the kind shown here. We can see that red stars of about the same luminosity as the Sun are to all intents and purposes absent. In type M we have either very large giants, or else very feeble dwarfs. There are no half-measures.

All this is bound up with the important problem of the source of a star's energy, and we must admit at once that things are not so simple as they might seem. It is untrue to say that the Sun and other stars are 'burning'. They are too hot to burn, but in any case there is an easy way of showing that they must draw their power from some other source. The first man to point this out was not an astronomer, but a Scottish physicist—William Thomson, afterwards Lord Kelvin, who lived from 1824 to 1907, and became perhaps the greatest British scientist of his time. Kelvin proved

THE HERTZSPRUNG-RUSSELL DIAGRAM. *This shows the relationship between a star's spectrum and its luminosity. Supergiants are on the uppermost part of the graph, the giant branch to the right, and White Dwarfs to the lower left. Most of the stars belong to the Main Sequence, which is shown by the white band. For many years it was supposed that this graph showed a strict evolutionary sequence, so that a star descended the Main Sequence from type B to M, but it is now known that matters are much more complex than this. The principle of graphs of this kind was due to the work of H. N. Russell, of the United States, and E. Hertzsprung of Denmark*

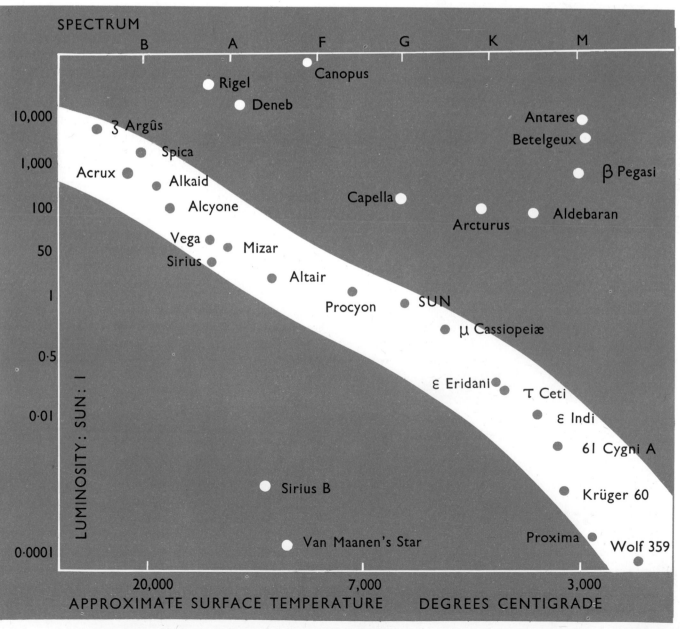

that if it were burning in the manner of a coal fire, the Sun would last only for a few thousands of years, whereas in fact the age of the Sun must be measured in thousands of millions of years.

Geologists can tell us the age of our own Earth with fair certainty. It proves to be about 4,700 million years, and the Sun is older still, so that the 'burning' theory is obviously wrong. (It is worth adding that men are relative newcomers to the scene. If we draw up a scale on which we represent the age of the Earth by eighty hours, the whole story of human civilization will have to be crammed into a single second.)

A better idea was put forward by Sir Norman Lockyer in 1890. Lockyer, the pioneer spectroscopist who paid such attention to the prominences of the Sun, was deeply concerned with the life-stories of the stars, and his plan of *stellar evolution* sounded delightfully straightforward. We know now that it was not correct, but nevertheless it provided a useful working basis.

Lockyer began by supposing that a star condensed out of material scattered in space as dust and gas. We know of such material; the Orion Nebula, close to the three bright stars of the Hunter's Belt, is made up of it, and this part of Lockyer's scheme is still accepted today.

Lockyer suggested that as the material drew together it would form a spherical mass, and its inner temperature would rise. Gravitational force would tend to make all the material condense towards a point in the centre of the mass, and so the temperature would increase steadily. At first the star would be a Red Giant of vast diameter, and relatively low surface temperature; as it shrank, it would become a Yellow Giant, and then a smaller but hotter bluish-white star of type B. This would be the peak of its career. The gravitational shrinking would go on, but the temperature would drop also, and the star would become successively a Yellow Dwarf (such as the Sun) and finally a Red Dwarf. Eventually all its heat would leave it, and it would become a cold, dead globe.

When Lockyer put forward this theory, the giant and dwarf divisions had not been recognized. The later work of Hertzsprung and Russell seemed to provide splendid confirmation, but still there was something wrong. A star which radiated only because it was shrinking under the influence of gravity could not last for more than 50 million years, which was not enough.

Russell tried to put matters right by introducing the idea of atomic energy. He knew of course that an atom contains protons, each of which carries a unit positive charge of electricity, and electrons, each of which carries a unit negative charge. If a proton and an electron met, he suggested, they would 'cancel each other out', and both would vanish, with the emission of energy. Could it be that a star shone because it was annihilating its material?

Russell's theory was published in 1913, and seemed to be satisfactory. A star would begin as a Red Giant, as Lockyer had supposed, and would turn into a hot white star, after which it would pass down the dwarf branch or *Main Sequence* until ending up as a Red Dwarf. Unfortunately there were difficulties of another sort. So much energy would be available that a star could last for at least 10 million million years, and this was as obviously

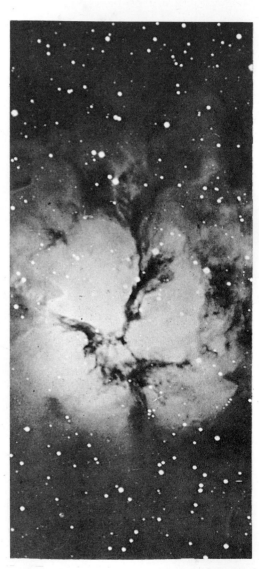

THE TRIFID NEBULA IN SAGIT-TARIUS, *photographed on June 30, 1921, with the 100-inch Hooker reflector at Mount Wilson; exposure 2½ hours. A nebula of this sort is a mass of dust and gas, and is entirely different from a galaxy such as the Andromeda Spiral, since it is a member of our own Galaxy. The old name for external galaxies, 'spiral nebulæ', is now obsolete. It was in any case somewhat misleading, since by no means all the galaxies are spiral in form*

too long as Lockyer's 50 million years had been too short. To make things worse, further studies of the atom showed that a proton and an electron could not annihilate each other as Russell had supposed.

Astronomers such as Sir Arthur Eddington, who had been born at Kendal, in Westmorland, in 1882 and who was responsible for great advances in our knowledge of a star's interior, puzzled over the whole problem of stellar evolution; yet almost up to the outbreak of the Second World War, the chief mystery—the source of stellar energy—remained unsolved.

The vital clue was discovered in 1938 by two astrophysicists working quite independently of each other, Hans Bethe in America and Carl von Weizsäcker in Germany. Apparently Bethe had been attending a conference in Washington, and was returning by train to Cornell University when he started wondering whether he could calculate any 'nuclear reactions' which would explain the reason why the Sun shone. Before he got out of the train, he had broken the back of the whole problem.

Hydrogen is much the most plentiful substance in the whole universe, and the Sun contains a tremendous amount of it. Near the Sun's centre, where the temperature is around 14,000,000 degrees and the pressure is colossal, strange things happen to the hydrogen nuclei. They band together to form nuclei of the second lightest element, helium. It takes four hydrogen nuclei to make one nucleus of helium, and the process is a complicated one, but the net result may be summed up quite simply: hydrogen is changed into helium, and energy is released in the process. It is this energy which is responsible for the Sun's radiation.

If it were possible to balance four hydrogen nuclei against the single helium nucleus formed from them, we would find that the helium is slightly the less massive. Therefore, mass has been lost in the energy production process. It appears that the Sun is losing mass by 4,000,000 tons every second. The figure seems

PROFESSOR SIR ARTHUR EDDINGTON. *One of the greatest of British astrophysicists, Sir Arthur Eddington was responsible for many important advances. He was moreover well known as a lecturer and broadcaster, and wrote a number of popular books on astronomy and allied subjects. Modern theories of stellar evolution owe much to his pioneer work in the earlier part of the present century*

STELLAR EVOLUTION ACCORDING TO LOCKYER. *In its final development, Lockyer's theory gave what appeared to be a very plausible explanation of stellar evolution. A star would begin its career as a Red Giant such as Betelgeux, pass down the giant branch until becoming a very luminous star of type B (Rigel), and then pass down the Main Sequence, becoming a Red Dwarf (Proxima) and, finally, a cold dark globe. It is now known that the whole theory is incorrect, and matters are in fact much less straightforward. Nevertheless, Lockyer's theory provided a useful working basis*

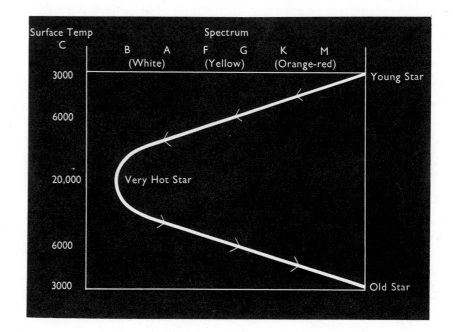

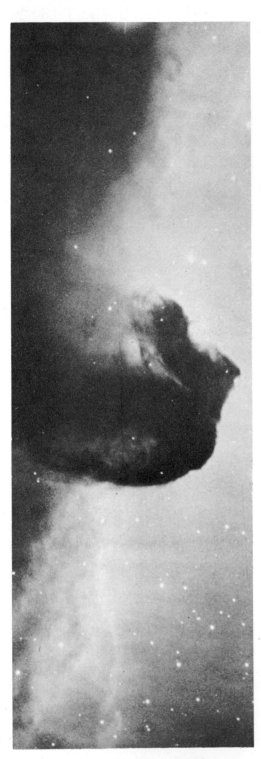

THE HORSE'S-HEAD NEBULA IN ORION, *near the bright star Zeta Orionis. This nebula, known officially as I.C. 434 or Barnard 33, is made up of dark material, and it is easy to see why it has been nicknamed the 'Horse's-Head'. The photograph was taken in red light with the 200-inch Hale reflector at Palomar*

unbelievable, but we may be sure that it is not very wide of the mark.

Hydrogen is the essential 'fuel', at least for stars such as the Sun, and of course the hydrogen supplies will not last for ever. Moreover, the very luminous stars are losing mass at an even more staggering rate. It has been estimated that Rigel in Orion does so at 80,000 million tons per second, and it cannot go on radiating so energetically for more than a few millions of years at most.

In view of all this, it is time to have another look at the Hertzsprung-Russell Diagram and to decide whether we can still retain the attractive Red Giant to Red Dwarf sequence. Rather reluctantly, astronomers have decided that we cannot. There is no choice but to start afresh.

We begin, as before, with a gaseous nebula such as the Sword of Orion. Local condensations appear, and the newly-born star contracts under gravitational force. As soon as the central temperature is high enough, nuclear reactions begin, and by the time the star has reached the Main Sequence we find that it is drawing its energy from the conversion of hydrogen into helium.

The main difference between modern and older theories is that a star such as the Sun is becoming more luminous as it uses up its hydrogen 'fuel'. We still do not know the full details, but it now seems certain that the energy output is becoming greater instead of smaller, so that the final fate of life on Earth will be destruction by heat instead of by cold.

Once a Main Sequence star has exhausted all the hydrogen available for 'burning', it must rearrange itself drastically. When the energy production has been halted, what seems to happen is that the core—now rich in helium—shrinks under the influence of gravitation. This causes a new rise in temperature, and at last new reactions begin, so that heavier elements are formed from the helium. A whole series of reactions takes place, and the inner temperature reaches fantastic values. Meantime the star's outer layers have been distended, and have become cooler; the star has become a Red Giant. Betelgeux and Antares are at this stage. Far from being youthful, they are approaching stellar senility. Our Sun will become a Red Giant eventually, with disastrous consequences for Earth, but we need not be alarmed. The change is not imminent, and the Sun will not leave the Main Sequence for at least 6,000 million years yet.

Actually it is not likely that the Sun will ever swell out to the size of Betelgeux or Antares, both of which had greater initial mass and are officially known as supergiants. Another Red Giant is Mira, the famous variable star in Cetus; many of these old stars fluctuate in light.

The giant or supergiant stage cannot last for ever. Heavy elements are created inside the core, but finally there comes a time when energy production cannot continue merely by element synthesis. This is where we are somewhat uncertain of the evolutionary sequence. A very massive supergiant may suffer a cataclysmic explosion, termed a supernova outburst, which destroys the star in its old form, and leaves nothing more than a mass of expanding gas together with a 'stellar remnant' of incredible density. Supernovæ will be discussed below. Meantime, it is

1844
1850
1860
1862
1870
1880
1890
1900
1910
1920
1930
1940
1950
1960

APPARENT MOVEMENT OF SIRIUS IN THE SKY. *The proper motion of the bright component, shown by the thick wavy line, showed that there must be an unseen companion affecting its motion*

probable that a more modest star, of solar type, will simply collapse into the kind of star known as a White Dwarf. This brings us on to the remarkable story of Sirius B.

Sirius itself is the most brilliant star in our skies. It is by no means particularly luminous; admittedly it is 26 times as powerful as the Sun, but this does not seem much when we remember that we know of stars which have a million Sun-power. Sirius is conspicuous mainly because it is only $8\frac{1}{2}$ light-years away. Of the 1st-magnitude stars, only Alpha Centauri, at $4\frac{1}{3}$ light-years, is nearer.

This closeness means that Sirius has an exceptionally rapid proper motion. In 1834, Friedrich Bessel realized that it was behaving in a peculiar manner. Its motion relative to its neighbors seemed to be erratic; instead of moving uniformly, it was 'waving' its way along, as shown in the diagram. Each 'wave' was very tiny, and took about fifty years to complete, but there seemed to be no doubt about it.

Bessel decided that there was only one explanation. Sirius must have a faint companion which was pulling on it and making it travel in a curve instead of a straight line. Just as Adams and Le Verrier later tracked down the planet Neptune because of its effects on Uranus, so Bessel worked out where the companion of Sirius must be. He had no hopes of seeing it himself; he realized that it must be overpowered by the glare of the bright star. When Bessel died, in 1844, the Companion was still undiscovered.

Then, in 1862, a well-known American instrument-maker named Clark decided to test a new 20-inch telescope by looking at Sirius. At once he saw a dim speck of light close beside the bright star. It proved to be the long-expected Companion, almost exactly where Bessel had said it would be.

According to its effects on Sirius itself, the Companion was estimated to be almost as massive as the Sun. Since it was only of magnitude $8\frac{1}{2}$, it was 10,000 times less luminous than its brilliant neighbor, and was assumed to be cool and red, with a large diameter.

The real shock came in 1915. At Mount Wilson, Walter Sydney Adams examined the spectrum of the Companion, and found to his surprise that the star was not cool at all. It was white, and its surface temperature was 8,000 degrees—2,000 degrees greater than the Sun's.

Astronomers all over the world were deeply interested. The Companion was very faint, so that it sent out relatively little light —and yet its surface was at white heat! The only way to make the observations fit was to assume that it was a dwarf with a diameter of only 24,000 miles. This meant that it was smaller than a planet such as Uranus or Neptune.

On the other hand the Companion was about as massive as the Sun, and so was quite unlike a planet. It was amazingly 'heavy' for its size, with a density of something like 70,000 times that of water. Comparing it with Sirius was rather like comparing a small, dense bullet with a distended cream puff.

Other 'White Dwarfs' were found, some of them even more extraordinary. One star studied by G. P. Kuiper in America proved to be only 4,000 miles across, about the size of Mars, and yet as massive as the Sun. If it were possible to take a cube of material from Kuiper's Star, with each side of the cube measuring one-tenth of an inch, and bring it back to Earth, the weight would be about half a ton. Many tons could be packed into a thimble.

To explain this 'super-dense' matter we must go back to our description of the way in which an atom is built up. There is a central nucleus, round which move circling or planetary electrons. (It is not easy to explain atomic structure in non-mathematical language, since the nucleus and the electrons are not, strictly speaking, solid lumps of material; but the picture is good enough for our present purpose.) It is clear, then, that most of the atom is empty space. High temperatures and pressures break up the atom by knocking off some of the planetary electrons; an atom damaged in this way is said to be *ionized*. In a White Dwarf, the ionization is complete. Every planetary electron is removed, and the result is a chaotic jumble of atomic nuclei and unattached electrons. Waste space is eliminated, and the broken pieces of the atom may be crammed close together. This accounts for the incredibly high densities.

SIZES OF DWARF STARS. *The Sun, with its diameter of 865,000 miles, is regarded as a dwarf star, but some of the White Dwarfs are much inferior to it in size. Here, the Sun is shown together with the planets Uranus, the Earth and Mars, and two White Dwarfs— Sirius B and Kuiper's Star. It is remarkable to find that Kuiper's Star, with a diameter slightly less than that of Mars, has a mass equal to that of the Sun, so that its matter is remarkably dense*

Uranus

Sirius B

There seems little doubt that a White Dwarf has used up all its nuclear fuel, and is continuing to shine only because it is still shrinking slowly under the influence of gravitation. It is a 'bankrupt' star, and has sunk into a long-drawn-out old age. No more supplies of energy are left to it, and eventually it must die.

Do all stars turn into White Dwarfs eventually? Some authorities believe so, in which case the life-story of a star of solar mass may be summed up quite neatly. The star begins by condensing out of interstellar material; it joins the Main Sequence, and may become very luminous; it turns into a Red Giant as its hydrogen is exhausted, and when it has used up its other fuels also it collapses into a White Dwarf. A more massive star may suffer a supernova outburst, and end its career as a pulsar.

The picture may be completely wrong. Theories change every few years, and new discoveries may cause us to alter our ideas completely. For the moment, however, we can say no more. There is at least a good chance that our Sun will experience a period of glory before it collapses and becomes a 'stellar glow-worm' similar to the strange, super-dense Companion of Sirius.

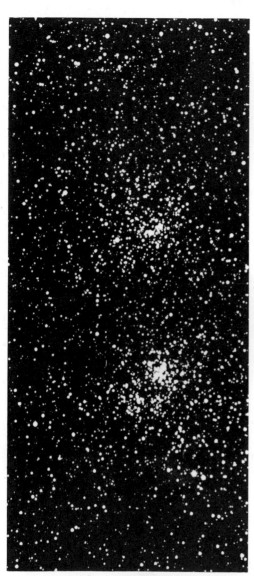

THE DOUBLE STAR-CLUSTER IN PERSEUS. *These objects are not included in Messier's famous catalogue, and are known officially as H.VI.33 and 34, from their numbers in Herschel's catalogue. The photograph given here was taken in 1932 by W. S. Franks, using a 6-inch refractor. The clusters, nicknamed the 'Sword-Handle' (not to be confused with the Sword of Orion) are beautiful objects, excellently seen in a small telescope*

WE HAVE SEEN how men such as Hipparchus, Ptolemy and Tycho Brahe drew up star catalogues, using observations made with the naked eye. The later application of telescopic sights allowed Flamsteed and his successors to draw up far better catalogues. In this connection it would be unfair not to mention Friedrich Argelander, who was born at Memel in Germany in 1799, and became Director of the Bonn Observatory. Between 1852 and 1863 he drew up a 'sky census' including all stars down to the 9th magnitude—a grand total of over 300,000. The *Bonn Durchmusterung* became a standard work.

Argelander was assisted by another German astronomer, Eduard Schönfeld, who later extended the catalogue to the southern skies and charted about 133,000 stars. Schönfeld became Director of the Bonn Observatory after Argelander died, and remained so until his own death in 1891.

Nowadays, of course, photography has enabled us to produce very detailed sky surveys. Particularly useful is the Schmidt camera, invented as recently as 1930, which has a rather complicated optical system which makes it possible to photograph large areas of the sky with each exposure. Large Schmidt cameras have been set up at Palomar and elsewhere. Bernhard Schmidt, who developed the principle, was originally interested in explosives, but having hurt himself so badly that he lost an arm he turned to astronomy as being rather safer. Astronomers throughout the world have every reason to be grateful that he did, since Schmidt cameras have proved to be of the greatest value.

Among the stars catalogued in early times were some which proved to behave in a curious way, since they did not shine steadily. Their brightness altered, sometimes to an extent of several magnitudes. Perhaps the most famous of these *variable stars* is Mira, in the constellation of Cetus, the sea-monster of the Perseus legend.

One of the early pioneers of telescopic astronomy was David Fabricius, a Dutch pastor who lived at Osteel in Holland. He was a friend of Tycho and Kepler, and father of Johann Fabricius, who discovered sunspots independently of Galileo and Scheiner but who died young.

On August 13, 1595, David Fabricius was looking at stars in Cetus when he noticed an object of the 3rd magnitude. He paid no particular attention to it, since it looked just like a normal star, but by October it had disappeared. It is rather curious that Fabricius made no attempt to follow the matter up. (Incidentally he met with an unfortunate end. In 1616 he preached a sermon in which he said that one of his geese had been stolen, and hinted that he knew who was responsible. Evidently he was right, since he was murdered before he could give the name of the thief.)

When Johann Bayer was drawing up his own catalogue, in 1603, he recorded a star in the same place as Fabricius had done. This time it was of the 4th magnitude, and Bayer gave it the

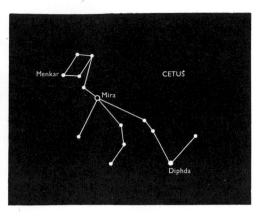

THE CONSTELLATION OF CETUS, *showing the position of the long-period variable Mira*

RICH STAR-FIELD IN THE MILKY WAY REGION

Greek letter Omicron. He did not associate it with Fabricius' vanishing star; neither did Wilhelm Schickard, Professor of Mathematics at Tübingen University, who saw it again in 1631.

Then, in 1638, another Dutchman—Johann Phocylides Holwarda, Professor at Franeker University—began a series of observations which showed that Omicron Ceti appeared and vanished regularly. It was a genuine variable star, and further observations by Hevelius from 1648 onward showed that it had a period of about 331 days. At maximum it may reach the 2nd magnitude, and has been known to outshine Polaris. At minimum it drops to magnitude $9\frac{1}{2}$, so that even binoculars will not show it. It has an M-type spectrum, and is strongly orange-red; like Betelgeux, it is a Red Giant of vast diameter. Moreover it has a faint companion which may be a White Dwarf.

Omicron Ceti was the first variable star to be identified, and was well named Mira, or 'the Wonderful'. Neither the period nor the maximum magnitude are constant. In some years Mira has failed to become brighter than the 5th magnitude, and on an average it is only visible to the naked eye for about 18 weeks out of its 47-week period. The last really bright maximum was that of the late summer of 1969, when for a time I ranked Mira only slightly inferior to the Pole Star.

Many other 'long-period variables' have been detected, a notable example being Chi Cygni in the Swan, which has a period of 409 days. At maximum it is of magnitude $4\frac{1}{2}$, and is an easy naked-eye object, but at minimum it sinks to magnitude 14, and a telescope of some size is needed to show it. Chi Cygni, too, has an M-type spectrum, and is a Red Giant. Most (though not all) of the long-period variables are of the same type. A small instrument will reveal many of them, and it is fascinating to watch their slow but obvious changes in brilliancy.

Rather different are the irregular variables, which have no definite periods at all. The most conspicuous of them is of course

HIGH-DISPERSION SPECTRUM OF MIRA CETI ▶

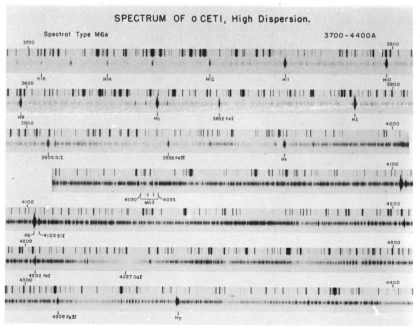

SPECTRUM OF o CETI, High Dispersion.

Spectral Type M6e

3700–4400A

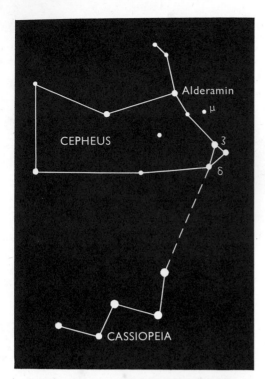

THE CONSTELLATION OF CEPHEUS, *showing the positions of the famous variables Delta and Mu. Cepheus is not a prominent group, but the W of Cassiopeia may be used as a pointer to it; its brightest star is Alderamin. Delta Cephei is not conspicuous, but is always an easy naked-eye object, and may be compared with its two neighbors Epsilon and Zeta, with which it forms a small triangle. Mu is sometimes easy to see without optical aid, but at its faintest it drops to almost the 6th magnitude. The strong red color which led to Herschel's christening it 'the Garnet Star' is not evident with the naked eye, but any small telescope will show it excellently*

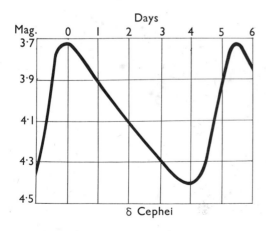

LIGHT-CURVE OF DELTA CEPHEI. *The fluctuations are perfectly regular. This star is the most famous member of its class*

Betelgeux in Orion, which sometimes matches Rigel, but may fade until it is no brighter than Aldebaran in the Bull. There is a rough period of about 5 years, but we can never tell just how Betelgeux is going to behave. This applies also to Mu Cephei, in the far northern part of the sky, which fluctuates between magnitudes 4 and 6. It is so highly-colored that Sir William Herschel nicknamed it 'the Garnet Star', and it is a lovely object in binoculars or a small telescope.

Some faint variables are even more unpredictable. R Coronæ Borealis in the Northern Crown is usually easy to see in a low-powered instrument, but sometimes it drops abruptly by as much as 8 magnitudes, remaining at minimum for some time before brightening up once more. Z Camelopardalis in the Giraffe normally shows semi-regular rises and falls, but sometimes becomes perfectly steady for months on end before starting to fluctuate again. Stars such as SS Cygni and U Geminorum have periodical 'outbursts', and are sometimes termed *dwarf novæ*. There seems to be no end to the different types.

Mention must also be made of an extraordinary star in the southern hemisphere, Eta Argûs. (Since Argo has been divided up, Eta Argûs has become Eta Carinæ.) It seems to be a variable and not a true nova, but its light-changes have been remarkable. About 1840 it was brighter than any star in the sky apart from Sirius; for the last half-century it has been too faint to be seen without a telescope. It is associated with the lovely gaseous Keyhole Nebula, and northern astronomers always regret that it never rises above the horizon in Europe or the Northern United States.

On the other hand there are some variable stars which are as regular as clockwork, so that we can always forecast their brightness with complete accuracy. The most famous of them is Delta Cephei, whose changes were first detected in 1784 by a most unusual astronomer—John Goodricke.

Goodricke was born in Holland, of English parents, in 1764. He was deaf and dumb, and remained so all through his life, but there was nothing the matter with either his eyesight or his brain, and he became an expert observer as well as a theorist. Unfortunately he lived only to the age of twenty-one. In spite of his handicap, he had shown every sign of becoming one of the leading astronomers of his time.

Goodricke was studying the stars in Cepheus when he was attracted by the changes of Delta, which forms a small triangle with its neighbors Epsilon and Zeta Cephei. There was a regular rise and fall, and soon Goodricke found that maxima occurred at intervals of 5 days 9 hours. The magnitude range was from 4·3, to 3·5, so that the star was always an easy naked-eye object even though it was never particularly conspicuous.

In the same year another English astronomer, Pigott, found that Eta Aquilæ, not far from the brilliant star Altair, varied from magnitude 3·7 to 4·5 in a period of just over 7 days. Like Delta Cephei, it was completely regular, and this applied also to many other variables found in later years, notably Zeta Geminorum in the Twins. Kappa Pavonis in the Peacock, which is too southerly to be seen from Britain or the Northern United States, is yet another similar variable.

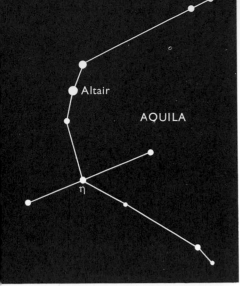

THE CONSTELLATION OF AQUILA, *showing the position of the Cepheid variable Eta Aquilæ*

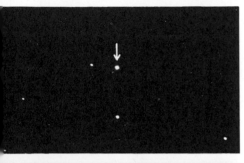

SUPERNOVA IN THE GALAXY I.C. 4168, *photographed with the 100-inch Hooker reflector at Mount Wilson. (Top) August 23, 1937, when the supernova was at maximum. (Middle) November 24, 1938, the supernova is now fainter. (Lower) January 19, 1942; the supernova can no longer be seen. Supernovæ of this sort have often been recorded in external galaxies*

Stars of such a kind are known as Cepheids, since Delta Cephei is the most celebrated member of the class. They have proved to be some of the most important stars known to us, because their periods are linked with their real luminosities; the longer the interval between one maximum and the next, the more luminous the Cepheid. For instance Delta Cephei, with a period of 5 days 9 hours, shines 660 times as brightly as the Sun when at its brightest. Eta Aquilæ, with a period of over 7 days, is more luminous. A southern Cepheid, *l* Carinæ, with a period of $35\frac{1}{2}$ days, has a maximum luminosity as great as 5,500 Suns.

The reason for this *period-luminosity law* is uncertain, but it is remarkably useful. As soon as we know the period of a Cepheid, which we can find out merely by studying it with a telescope, we can tell its luminosity. By comparing the luminosity with the apparent magnitude, we can work out the star's distance. Since Cepheids are powerful stars, they remain visible over vast distances, and have been named the 'standard candles' of the Galaxy. By using them we can gauge distances which are so great that the old parallax method would be utterly useless.

It is interesting to find that the Pole Star is a Cepheid variable. Its changes in magnitude are very slight, and its period is almost 4 days, which tells us that its luminosity is about 550 times that of the Sun.

Related to the Cepheids are the RR Lyræ variables, which are just as regular but which have shorter periods. All are remote, and so all appear faint. The strange thing about them is that all seem to have about the same luminosity, 90 times as great as the Sun's. They are so called because RR Lyræ, in the small but interesting constellation of Lyra (the Lyre) was the first to be discovered.

Associated with variable stars are the novæ, once called temporary stars or new stars. As we have seen, they are not genuinely 'new' at all, but represent some violent outburst in a previously faint star. With a normal nova, the outburst affects only the outer layers, and after its blaze of glory the star returns to something like its old state. There are even a few stars which have been known to undergo more than one outburst. Such is T Coronæ, in the Northern Crown, which is usually of the tenth magnitude, but which rose abruptly to the second magnitude in 1866 and again in 1946. Whether these 'recurrent novæ' should be classed as of true nova type or as curious irregular variables remains uncertain. Eta Argûs (Eta Carinæ) may be a recurrent nova; there is also P Cygni, an extremely hot, remote, luminous and unstable star in the constellation of the Swan, which reached the third magnitude in 1600, but is now only dimly visible with the naked eye.

The most brilliant novæ of modern times have been the stars which appeared in Perseus (1901) and in Aquila (1918), which outshone all the stars apart from Sirius and Canopus. Nova Persei later developed a 'shell' of gas which showed many interesting features; so did Nova Aquilæ, though in this case the gas has now become too dim to be visible. Gaseous material was also ejected by Nova Lacertæ of 1936. The material was sent out at velocities of something like 2,400 miles per second.

One of the most interesting of modern novæ was discovered in December 1934 by J. P. M. Prentice, an English amateur famous

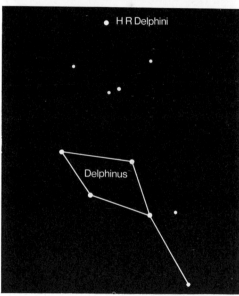

for his work in connection with meteors. It lay in Hercules, no far from Vega and the Dragon's head. It reached the first magnitude before fading away again, and is still visible in powerfu telescopes; today it is an eclipsing binary system, but there ar intrinsic fluctuations as well. At maximum, I remember seeing it almost as bright as Altair.

Another English amateur, G. E. D. Alcock of Peterborough has the splendid record of having discovered three novæ (as wel as four comets). The first and most interesting of these was Nova Delphini of 1967, now officially called HR Delphini. When Alcock found it, during a routine binocular 'sweep' over the area o Delphinus (the Dolphin) it was of about the fifth magnitude, bu unlike most novæ it did not fade quickly; it reached peak brilliance late in the year, and then began a very slow decline. Not until 197 did it become too faint to be followed with binoculars.

Fortunately we have some knowledge of HR Delphini in the pre-outburst stage. It was of the twelfth magnitude, and had an O-type spectrum. Whether or not it will fade below its origina magnitude remains to be seen, but it is unlikely to disappear from view, and it will probably settle down to its old obscurity. It i almost 3,000 light-years away, so that the actual outburst tool place around the year 2000 B.C. This is much further away than Prentice's nova in Hercules, which lies at 800 light-years, and actually flared up in the reign of William the Conqueror's son Henry I!

Alcock's second nova, found in Vulpecula in April 1968, wa more conventional. From the fifth magnitude at its discovery i plunged to the tenth by the end of the year, and has now become very faint. And his third discovery, in Scutum (the Shield) in 1970, was also a nova of the fast type; it never attained naked-eye visibility.

Nova-hunting is an attractive pastime for the dedicated amateur but it is laborious and time-consuming, and the enthusiast may well go through his lifetime without making a discovery. Yet a new nova may flare up at any moment; we cannot tell.

Supernovæ are much more violent, and represent stellar deaths Tycho's star of 1572 has already been described. Kepler's star o 1604, in Ophiuchus, was also a supernova. But the greatest fasci

HR DELPHINI (upper), photographed by Commander H. R. Hatfield in 1967. The nova was then visible with the naked eye, but is now too faint to be seen without a telescope. POSITION OF HR DELPHINI *(lower), in the region of Delphinus*

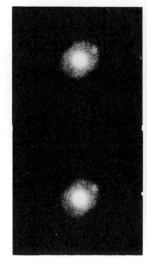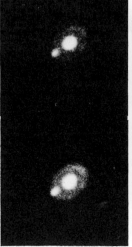

NOVA HERCULIS 1934, *photographed in ultra-violet, green and red light in 1951. Photographs with the 200-inch Hale reflector at Palomar*

THE KEYHOLE NEBULA IN ARGO.
*Nebulosity in the region of the famous
irregular variable Eta Argûs. Unfortunately
the area is too far south in the sky to be seen
from Europe or the northern United States*

nation is that of the supernova of 1054, because its remnants are still visible with a modest telescope in the form of the gas-cloud which we call the Crab Nebula.

The Crab lies near the third-magnitude star Zeta Tauri, and to the casual observer looks like a filmy patch. Photographs taken with giant telescopes show its intricate structure, and we can appreciate why the nickname of 'Crab' was bestowed on it more than a century ago by the third Earl of Rosse. Officially it is termed Messier 1, since it was the first entry in Charles Messier's celebrated 1781 catalogue of clusters and nebulæ.

There is no doubt whatsoever that the Crab is the débris of the star of 1054. Its distance from us is 6,000 light-years, and the rate of expansion is over 700 miles per second, so that each day its diameter increases by about 70,000,000 miles. To modern astronomers it is one of the most significant objects in the sky, because it sends us not only visible light but also radio waves, X-rays and other short-wave radiations. In our experience it is unique, and it has added greatly to our knowledge. One astronomer has been quoted as saying that there are two branches of astronomy: the astronomy of the Crab Nebula, and the astronomy of everything else!

During the past few years another remarkable feature of the Crab has been discovered. It contains one of the very small, rapidly-fluctuating radio sources known as pulsars, now believed to be neutron stars in which even the nuclei of atoms are crushed together to make material of incredible density—far surpassing even that of a White Dwarf. The pulsar in the Crab has been identified with a very faint flashing object, and remains the only pulsar to be linked with something which can actually be seen. Inevitably it has led to the suggestion that all pulsars are the end-products of supernovæ, and this may well be the case, though it is too early to be sure. At any rate, the Crab pulsar is an amazing object. A cupful of its material would weigh millions of tons.

Supernovæ have been seen in other galaxies; one of them, S Andromedæ of 1885, flared out in the Andromeda Spiral, and reached the fringe of naked-eye visibility even though we saw it across a distance of more than two million light-years. We cannot forecast when the next supernova in our own Galaxy will appear. Astronomers would very much like to study a galactic supernova with the aid of modern equipment (the last, Kepler's Star of 1604,

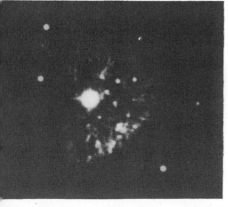

EXPANDING NEBULOSITY ROUND
NOVA PERSEI, *photographed with the 200-
inch Hale reflector at Palomar*

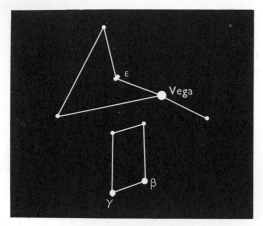

THE CONSTELLATION OF LYRA

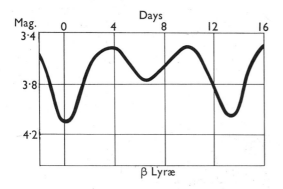

β Lyræ

LIGHT-CURVE OF BETA LYRÆ. *The secondary minimum is much more pronounced than in the case of Algol*

occurred in pre-telescopic times), but undoubtedly a supernova is something to be observed from a respectful distance.

In 1669 Geminano Montanari, Professor of Mathematics at the Italian University of Bologna, found that Algol (Beta Persei) was apparently variable. At this time Mira was the only variable star known, so that Montanari's discovery caused great interest, particularly as Algol behaved in a way of its own. Usually it was steady at just below the 2nd magnitude, and showed no change for a total period of 2 days 11 hours. Then it faded down to magnitude $3\frac{1}{2}$, taking 5 hours to do so; it remained faint for a short period, and then took a further 5 hours to climb back to maximum.

Goodricke, painstaking as ever, studied Algol closely and came to the conclusion that the star was not a genuine variable at all. Admittedly it showed fluctuations in light, but Goodricke believed that these changes were due to some darker body passing between Algol and ourselves. In this case Algol would be a binary, with one bright and one dimmer component; when the dimmer star passed between the bright component and ourselves, it would eclipse the main star, and the total brightness would drop.

As we now know, Goodricke was quite right. The bright member of the Algol pair has a B8-type spectrum and a surface temperature of 12,000 degrees, with a diameter of about $2\frac{1}{2}$ million miles. The faint component is larger but less brilliant. When the bright star eclipses the faint star there is a much smaller fall in brilliancy, so slight that even Goodricke did not notice it.

Strictly speaking, then, Algol is not a proper variable, but an eclipsing binary. There is no difference between it and an ordinary binary of short period except that the plane of the orbit happens to lie in our line of sight. Were the orbit differently tilted, no eclipse would occur, and Algol would shine steadily.

Another eclipsing variable easily visible to the naked eye is Beta Lyræ, close to the brilliant Vega. Here the light-changes are always going on. Both components are very hot and luminous, and are so close together that they almost touch. They raise huge tides in each other, and each star is distorted into the shape of an egg; they draw material out of each other, and some of this material streams out to produce a shell of gas round the pair. Fewer than 200 Beta Lyræ-type variables are known, but eclipsing binaries similar to Algol are much more common.

Perhaps the most remarkable of the eclipsing binaries are Zeta and Epsilon Aurigæ, which with Eta Aurigæ form a triangle of stars beside Capella. Zeta consists of a B8 Main Sequence star over 100 times as luminous as the Sun, accompanied by a K4-type supergiant with a much lower surface temperature but with a diameter of at least 200 million miles. The period is 972 days, as against less than 4 days for Algol, which shows that the components are much farther apart. When the red star eclipses the hot giant, interesting changes in the spectrum are noted; the hot star seems to dim slightly, showing that it is shining through the vast tenuous atmosphere of its companion, and it is some time before the spectrum of the hot star disappears. From this we can calculate that the outer part of the red component has a density of no more than about one-five-millionth that of water.

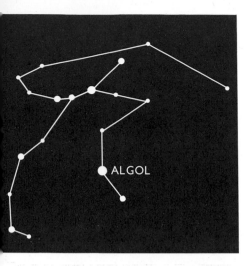

THE CONSTELLATION OF PERSEUS

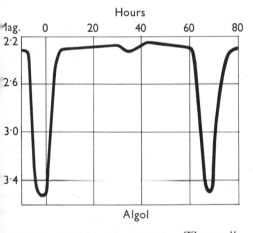

LIGHT-CURVE OF ALGOL. *The small secondary minimum cannot be detected without instruments*

Epsilon Aurigæ is even more remarkable. It has always been regarded as a very youthful eclipsing binary, with an F-type primary 60,000 times as luminous as the Sun; to the naked eye it appears as an ordinary star, and even telescopes will not show the secondary, which radiates only in the infra-red. Every 27 years the primary suffers eclipse, and the magnitude drops from 3 to 4. Because a complete eclipse sequence lasts for over 700 days, it has been calculated that the invisible secondary must have a diameter of about 1,800 million miles—so that it could contain the orbits of the planets in the Solar System out to and beyond Uranus. The masses have been estimated at 35 Suns for the primary and 23 Suns for the secondary.

On this picture, the secondary is a 'proto-star', still shrinking as it forms from the interstellar material, and not yet on the Main Sequence, so that the hydrogen-into-helium process has not begun. A vehicle moving round it at a steady 1,000 m.p.h. would need over 6,000 years to make one circuit, and this would make it the largest individual star known to science.

In 1971 this explanation was challenged by A. G. W. Cameron and R. Stothers, of the Goddard Institute for Space Studies in New York. To them, the system of Epsilon Aurigæ is not young, but is very well advanced in its evolution. The bright primary has left the Main Sequence, after the exhaustion of its available hydrogen, and is now 'burning' helium, so that heavier elements are being synthesized inside it. The secondary is even more advanced. It has suffered some vast explosion (probably a supernova outburst) and has then undergone an 'implosion'. Its material has been drawn together by violent gravitational collapse and by now no light or material can escape from its area, so that it has become what may be termed a 'black hole' in the Galaxy—though Cameron's more scientific name of 'collapsar' will certainly become official. The infra-red radiation we can detect comes from small particles orbiting the collapsar at about 15,000 million miles from the centre of mass of the object. Once they spiral inward, and are captured by the collapsar, they too will vanish from our ken.

Certainly this is a strange concept, but collapsars have long been known to be theoretical possibilities, as all relativity theorists agree. So far as Epsilon Aurigæ is concerned the evidence is very incomplete, and the system may after all prove to be an ordinary eclipsing binary. Yet in either case it is of exceptional interest. Its variability has been known for over a century and a half (it was detected by Fritsch in 1821) but only in modern times has its remarkable nature become apparent.

Eclipsing binaries prove that some apparently single stars are in fact pairs, with components so close together that no telescope will divide them. However, the German astronomer Vogel, who acted as Director of Potsdam Observatory between 1882 and his death in 1907, found that where the telescope fails the spectroscope can succeed. We come back once more to our reliable ally the Doppler Effect.

Suppose that we have a very close binary whose components appear as a single mass? If the orbital plane is almost edge-on to us, as shown in the diagram, there will be times when one star (A)

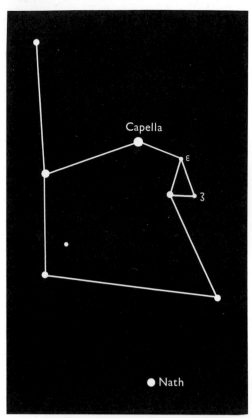

THE CONSTELLATION OF AURIGA,
*showing the positions of the two remarkable
eclipsing binaries Epsilon and Zeta Aurigæ*

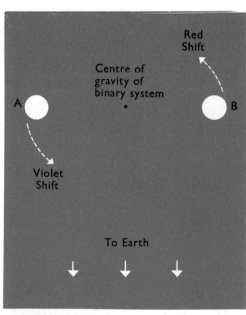

DETECTION OF A SPECTROSCOPIC
BINARY. *In the positions shown, component
A is approaching and will show a violet shift;
B is receding, and will show a red shift*

is approaching us, while the other (B) is receding. The spectrum of A will therefore yield a Doppler shift towards the violet, while the spectrum of B will give a red shift. At other times, when A is receding and B approaching, the shifts will be reversed; when the motion of the components is transverse to the line of sight, there will be no Doppler shifts due to orbital motion.

If therefore we find a star whose spectrum lines regularly become double, we may be sure that we are dealing with a close *spectroscopic binary*. Capella is one such star, while the bright component of Mizar in the Great Bear is another.

Visual binaries are much more common than optical pairs of the kind shown in the diagram on page 97. Some of them show contrasting colors, and are really beautiful. Perhaps the loveliest of all is Albireo in the Swan, which has a yellow primary and a bluish-green companion and is well seen in a small telescope. Antares, the bright red star in the Scorpion, has a green companion; so has Rasalgethi in Hercules, which is one of the largest stars known to us and is an excellent example of a red supergiant.

The stars of a binary system move round their common centre of gravity. The periods range from a few hours for close spectroscopic pairs up to millions of years for wide visual pairs. As we have seen, the stars are not so unequal in weight as might be supposed; even Mira Ceti, with its vast diameter of about 200 million miles, has less than 20 times the mass of the Sun. Therefore the centre of gravity for a normal binary system is not very far from the mid-point of an imaginary line joining the centres of the two stars.

It used to be thought that a binary must be the result of the fission of a formerly single star, which broke up because of over-rapid rotation, but this view is not now generally held, and most astronomers believe that the components of a binary were formed at around the same time from the same cloud of interstellar material.

Multiple stars also occur. The most spectacular is Epsilon Lyræ, near Vega. To the naked eye it appears as a double, and a 3-inch refractor is enough to show that each component is again double, so that we have a quadruple system. Castor, the senior though fainter member of the Heavenly Twins, is a six-fold system made up of four bright components and two dim stars. And in the Orion Nebula there is the magnificent multiple Theta Orionis, nicknamed the Trapezium for reasons which are obvious to anyone who has seen it.

Only a few of the most interesting objects in the stellar heavens have been mentioned here. There is endless variety among the stars, and our knowledge is increasing all the time. The science of astrophysics came into its own only about a century ago, but it has revolutionized our entire outlook. The last few decades have been particularly productive. Even if we are still ignorant about the fundamental nature of the universe, we have learned a great deal about the life-stories of the stars themselves.

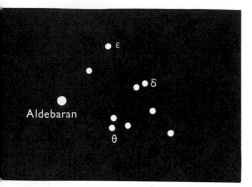

THE HYADES, *which appear as a V-shape of relatively faint stars near the brilliant red Aldebaran, in Taurus (the Bull)*

25   the galaxies

LOOK UP AT THE SKY on a clear night, and you may think that you will see millions of stars. This is not the case. At any one moment, nobody can see more than a few thousands of stars without using optical aid. However, the actual number of stars in our Galaxy really does amount to millions; indeed, the best modern estimate gives a grand total of 100 thousand million.

Some of these stars are collected together in *open* or *galactic clusters*, of which the best examples are the Pleiades or Seven Sisters, round the 3rd-magnitude star Alcyone in Taurus, and the Hyades, round Aldebaran—also in Taurus. Another interesting cluster visible to the naked eye is Præsepe in Cancer, nicknamed the 'Beehive', while southern observers have the 'Jewel Box' in the Southern Cross.

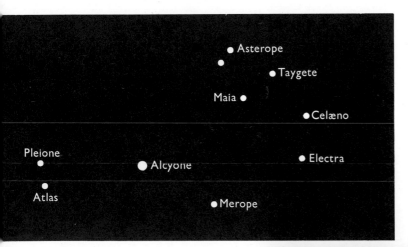

THE PLEIADES *or 'Seven Sisters'. All the stars named are visible with the naked eye on a clear night. The brightest of them, Alcyone, is of the 3rd magnitude*

SPIRAL GALAXY N.G.C. 2841, IN URSA MAJOR. *The spiral form is well shown. Like most external galaxies, the object is not well seen with a small or moderate telescope, and photography with very large instruments is needed to bring out the details properly. This picture was taken with the 200-inch Hale reflector at Palomar*

SPIRAL GALAXY MESSIER 64 (N.G.C. 4826) IN COMA BERENICES,
as photographed with the 60-inch reflector at Mount Wilson

NGC 205, *companion of the Andromeda Nebula, photographed in red light with the 200-inch Hale reflector at Palomar*

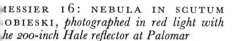

MESSIER 16: NEBULA IN SCUTUM SOBIESKI, *photographed in red light with the 200-inch Hale reflector at Palomar*

There are also the nebulæ, which are less conspicuous. The first mention of them goes back to the tenth century, when Al-Sûfi recorded the Great Spiral in Andromeda, though to him it appeared only as a faint misty patch. The Orion Nebula was mentioned in 1612 by Peiresc, and the work of Messier, Herschel and others led to the detection of large numbers of nebulæ of all kinds. Herschel, as we have seen, regarded some of them as 'starry' and others as non-stellar. In 1791, writing about the Orion Nebula, he stated: 'Our judgment, I venture to say, will be that the nebulosity around the star is not of a starry nature.'

Once again the decisive experiment had to wait until the development of the spectroscope. It was made on August 29, 1864, by Sir William Huggins.

A star yields a spectrum with a rainbow background and dark absorption lines. If therefore a nebula is made up of stars, the result will be a jumble of all the separate star-spectra together, and the main absorption lines should be made out. If a nebula is made up of gas, it will give an emission spectrum consisting of isolated bright lines.

Huggins turned his spectroscope towards a 'planetary' nebula in Draco. At once he saw bright emission lines, but no rainbow background. Herschel had been right; the nebula was gaseous, and was not made up of stars.

Irresolvable nebulæ are known as *gaseous* or *galactic nebulæ*.

THE DUMB-BELL NEBULA IN VUL-
PECULA, *a famous planetary, photographed at the Lick Observatory*

PLANETARY NEBULA N.G.C. 7293
IN AQUARIUS, *photographed in red light
with the 200-inch Hale reflector at Palomar*

The Sword of Orion is the best known of them; it surrounds a famous multiple star, Theta Orionis (known as the Trapezium, because of the arrangement of its four main components), and is of tremendous extent. A ray of light would take twenty years to cross it from one side to the other. It is in nebulæ of this sort that, according to modern theory, fresh stars are being created out of the dust and gas.

Herschel also noted various different objects which he christened *planetary nebulæ*, because they showed planet-like disks. The brightest of them, the Ring Nebula, is conveniently placed between the famous eclipsing binary Beta Lyræ and its neighbor Gamma, and is easy to see in a moderate telescope. Actually a planetary nebula is neither a planet nor a nebula. It consists of a faint central star, possibly not unlike a White Dwarf, surrounded by a tremendous, expanding shell of gas. Some authorities believe that a planetary nebula represents the final stage of a nova or supernova, but of this there is no proof. The gas is very rarefied; if we could take a cupful of air and spread it around a vast enclosure 5 miles in diameter, the resulting density would be about the same as that of a planetary nebula.

All Herschel's most important work was concerned with the distribution of stars in our Galaxy. He believed the system to be shaped like a double-convex lens, with the Sun near the middle,

MESSIER 13, THE GLOBULAR CLUSTER IN HERCULES, *photographed with the 60-inch reflector at Mount Wilson. The first photograph was given an exposure of 15 minutes; the second, 37; the third, 94. The increase in the number of stars shown with increasing exposure is very obvious. In this respect the photographic plate is far superior to the eye; the longer the exposure, the more it records*

THE OWL NEBULA, *Messier 97 in Ursa Major. Another planetary, photographed with the 60-inch reflector at Mount Wilson*

and at the time his idea seemed perfectly reasonable. It was not until well into the twentieth century that a more accurate picture was given by the American astronomer Harlow Shapley.

Shapley—who is still alive and active—was born at Nashville, in Missouri, in 1885. After completing his university education he worked at Princeton Observatory under H. N. Russell, of Russell Diagram fame. In 1914 he moved to Mount Wilson, and stayed there until 1921, when he was appointed Director of Harvard College Observatory. He continued to work as hard as ever even after he officially retired.

While at Mount Wilson, Shapley used the 60-inch reflector there to study the objects known as globular clusters. These are immense 'balls' of stars, comparatively crowded towards the centre of the clusters, and are magnificent in large instruments. About 100 of them are known. Messier 13, in Hercules, is visible to the naked eye, and is a familiar sight to European astronomers, but the brightest globular clusters—Omega Centauri and 47 Tucanæ—lie too far south to be seen from our latitudes.

Shapley knew that the globulars are very distant, and seem to form a kind of 'outer framework' to the Galaxy. They were so remote that they showed no measurable parallaxes, and Shapley turned to less direct methods. Moreover he saw that most of the globulars were in the southern sky, particularly in the region of the constellations Scorpio and Sagittarius. Such a 'lop-sided' distribution could hardly be due to chance.

The essential clue was found when Shapley detected RR Lyræ

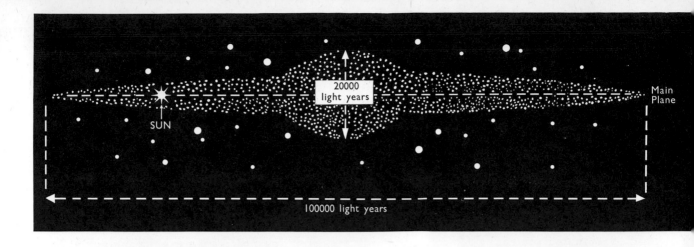

20000 light years

100000 light years

Main Plane

SUN

THE GLOBULAR CLUSTER OMEGA CENTAURI, *photographed at Royal Observatory, Cape, with the 24-inch Victoria telescope on February 19, 1903 (exposure 50 minutes). This is the brightest globular cluster in the sky but unfortunately it lies too far south in the sky to be seen from Europe or the northern United States*

GALAXY N.G.C. 4565, *a system seen almost 'edge-on' to us*

variables inside the globular clusters. It was known that these stars obeyed a period-luminosity law, and that each was about ninety times as luminous as the Sun. Once RR Lyræ variables had been found, then, their distances could be worked out—and hence, the distances of the globulars in which they lay.

Shapley's results gave astronomers something of a shock. By studying the globulars and measuring their distances it has been shown that the Galaxy has a diameter of about 100,000 light-years with a greatest width of about 20,000 light-years. The Sun lies well away from the middle, and is in fact about 32,000 light-years from the *galactic nucleus,* which at once explains the lop-sided distribution of the globular clusters. The galactic nucleus lies in the direction of the glorious rich star-clouds in Sagittarius.

Unfortunately we cannot see right 'through' the nucleus, or even to the galactic centre. There is a great deal of obscuring material in the way, and light-waves cannot penetrate it. Radio waves can do so; but when Shapley announced his results, soon after the end of the First World War, radio astronomy still lay in the future.

Following on from this research, it was found that the whole Galaxy is rotating. The Sun is moving round the nucleus, taking over 225,000,000 years to complete one journey. This period is sometimes termed the *cosmic year.* One cosmic year ago, the most advanced life-forms on Earth were amphibians; the Coal Measures were being laid down, and mammals had not yet appeared. Two cosmic years ago, the only life on Earth consisted of tiny creatures in the oceans. It is interesting to speculate as to what 'men' will be like in one cosmic year from now—if, indeed, men still exist.

Now let us go back to Herschel's second guess: that the starry or resolvable nebulæ were 'island universes', far beyond the boundaries of our own system.

The clusters, as well as the gaseous nebulæ such as the Sword of Orion and all the stars visible to the naked eye, were known to be members of our Galaxy, but it was difficult to be so sure about the spirals and other objects of similar type. Were they, too, members of one large system—or were they not? The only way to decide was to measure the distance of at least one of them, and the most promising was undoubtedly the Great Spiral in Andromeda, Messier 31. Since it was bright enough to be seen without a telescope, it was likely to be nearer than the rest.

The matter was settled in 1923 by Edwin P. Hubble, who had been assistant at Yerkes Observatory from 1914 to 1917, and had

DIAGRAM OF THE GALAXY AS IT WOULD BE SEEN EDGE-ON. *The galactic centre lies about 25,000 light-years away from us. The Sun is therefore well out toward the edge of the system, and is not near the centre, as Herschel believed. When we look along the main plane, many stars are seen in approximately the same direction, and this causes the Milky Way effect*

THE 'SATURN GALAXY' N.G.C. 4594, *photographed with the 60-inch reflector at Mount Wilson. The nickname is due to a superficial resemblance to Saturn and its system of rings*

MESSIER 87, *a globular galaxy in Virgo, photographed with the 200-inch Hale reflector at Palomar*

then moved to Mount Wilson. Using the 100-inch Hooker reflector, much the most powerful telescope in the world at that time, he searched for Cepheid variables inside the Andromeda Spiral, and finally he found them. He identified half a dozen, and worked out their distances. The result was conclusive. The Cepheids, and hence the Spiral, were hundreds of thousands of light-years away, and so the Spiral was in fact a separate system, containing stars, clusters, gaseous nebulæ and other features known in our Galaxy.

A new system of naming came into force. The term 'nebulæ' was confined to the gaseous objects in our stellar system, while the spirals and other resolvable objects became known simply as 'galaxies'.

Originally, Hubble estimated the distance of the Andromeda Galaxy as 750,000 light-years. He later increased this to 900,000 light-years, but in September 1952 astronomers had yet another shock. W. Baade, using the 200-inch Palomar reflector, announced that there had been a mistake in the Cepheid scale, and that all the galaxies were at least twice as remote as had been thought. The distance of the Andromeda Spiral was not 900,000 light-years, but over *two million*. This meant that it was a system even larger than the Galaxy in which we live.

Years before, Baade himself had pointed out that there seem to be two kinds of 'star populations'. The brightest stars of Population I are very luminous, and are of spectral types O and B; in these regions there is considerable dust and gas spread through space. With Population II the brightest stars are red supergiants, and there is almost no interstellar material. Globular clusters and the centres of galaxies are mainly Population II, while the spiral arms of galaxies are mainly Population I.

When mapping the Galaxy, Shapley had used RR Lyræ stars, not 'classical' Cepheids; and his results were correct, since his estimates of the luminosities of RR Lyræ stars were valid. What neither he nor anyone else had realized was that there are two types of short-period variables; those of Population I (including RR Lyræ stars) and those of Population II (including the classical

STELLAR POPULATIONS I AND II. *The Andromeda Galaxy, photographed in blue light, shows giant and supergiant stars as Population I in the spiral arms. The hazy patch at the upper left is composed of unresolved Population II stars. N.G.C. 205, companion of the Andromeda Galaxy, photographed in yellow light, shows stars of Population II; the brightest stars are reddish, and only 1/100 as bright as the blue giants of Population I. The very bright, uniformly distributed stars in both pictures are foreground stars belonging to our own Galaxy. Photographed with the 200-inch Hale reflector at Palomar*

NUCLEUS OF THE ANDROMEDA GALAXY, *showing Population II stars resolved. Photographed with the 200-inch Hale reflector at Palomar*

Cepheids), and there is a great difference between them. A Population I variable is much more luminous than a Population II variable of the same period.

The Cepheids used by Hubble to measure the distance of the Andromeda Spiral were of Type I, but the distances had been worked out on the assumption that they were of Type II. Therefore since the Cepheids were more luminous than had been thought, they must also be more remote; and this led to a

complete revision of our ideas of the scale of the universe. It also explained why RR Lyræ stars did not appear in the Andromeda Spiral. Undoubtedly they existed there, but they are less powerful than the classical Cepheids, and were too faint to be seen even with the Palomar reflector.

Not all the outer galaxies are spiral, and in fact spirals seem to be in the minority. Some, such as the two Nubeculæ or Magellanic Clouds, are more or less irregular (though it has been suggested that there are vague indications of spiral structure). The Nubeculæ, are too far south to be seen from Europe, or the Northern United States, and are the closest of our neighbor systems. They lie about 180,000 light-years from us. They seem to be 'satellites' of our Galaxy, and are magnificent objects, very conspicuous to the naked eye. The Large Cloud (Nubecula Major) is particularly splendid, and contains stars of high luminosity. One of them, S Doradus, is thought to be about a million times as powerful as the Sun—and yet from Earth it cannot be seen at all without optical aid.

We also meet with elliptical galaxies, which look at first sight not unlike globular clusters, though they are of course entirely different in nature. Then there are the extraordinary barred spirals, as well as the spectacular 'Catherine wheels' such as Messier 51 in Canes Venatici (the Hunting Dogs). The only way to study these various objects properly is to take photographs of them with large telescopes, and it is true to say that even the Andromeda Galaxy is rather disappointing when observed with a modest instrument; it looks like nothing more than a fuzzy patch. However, pictures taken with the great reflectors show the galaxies in all their glory.

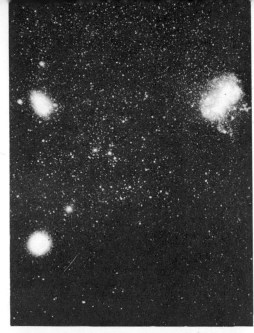

THE NUBECULÆ OR MAGELLANIC CLOUDS. *The Large Cloud is to the right, the Small Cloud upper left. The globular cluster in Tucana is also shown* (lower left)

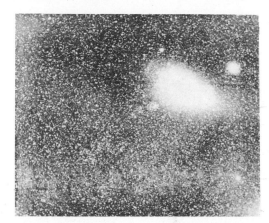

THE NUBECULA MINOR, *or Small Magellanic Cloud*

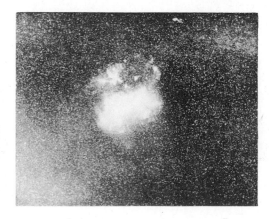

THE NUBECULA MAJOR, *or Large Magellanic Cloud*

THE WHIRLPOOL GALAXY, *Messier 51 in Canes Venatici, photographed by Ritchey with the 60-inch reflector at Mount Wilson*

SPIRAL GALAXY IN LEO, N.G.C.
2903, *photographed with the 200-inch Hale
reflector at Palomar*

CLUSTER OF GALAXIES IN COMA
BERENICES, *photographed with the 200-
inch Hale reflector al Palomar. The cluster
is about 40,000,000 light-years away*

Another important discovery was that the galaxies tend to collect in clusters. Among the members of our local cluster are the Galaxy in which we live, the Nubeculæ, the Andromeda Spiral, the fainter spiral in Triangulum, and more than a dozen other systems. Beyond, we have to travel an immense distance before we come to the next group of galaxies.

Since the members of the local group are comparatively near on the cosmical scale, the Cepheids in them show up plainly. As we probe farther and farther into space, the Cepheids merge into the general background blur, but supergiants are still visible, and by taking an average luminosity for a supergiant star we can estimate the distances of galaxies as far out as 20 to 25 million light-years.

In the constellation of Virgo lies a whole group of galaxies known as the Virgo Cluster (not really a happy term, since a cluster of galaxies is in no way related to a star-cluster such as the Pleiades or the Hercules globular). By observing the supergiants in the Virgo galaxies, we can work out the distances of the galaxies themselves and decide how large they are. From this, it is possible to estimate the average size of galaxies of different kinds. Beyond about 25 million light-years we lose even the supergiants, but if we know the shape of a more remote galaxy we can get some idea of its distance—though probably not with much accuracy. The most distant galaxy yet studied optically, 3C–295 in Boötes, is about 5,000 million light-years away.

Five thousand million light-years! We are indeed looking through both space and time. When we study the Boötes galaxy, we are seeing it not as it is now, but as it used to be 5,000 million years ago, before the Earth existed in its present form. And even the Boötes galaxy does not represent the limit of our investigations. Quasars, to be described below, seem to be more remote still.

On a photograph the remote galaxies appear as tiny, blurred patches, and it is difficult to realize that each is a tremendous system containing at least 100 thousand million stars, many of

THE VEIL NEBULA IN CYGNUS. *The
colors radiated by these filaments of gas result
from their motion through space. Ejected from
an exploding star more than 50,000 years ago
with an initial velocity of nearly 5,000 miles
per second, the clouds of gas have now been
slowed to a speed of about 75 miles per
second by constant collisions with atoms in
interstellar space. The force of these collisions
ionizes the gas and causes it to glow with the
characteristic colors shown here. Clouds
ejected in the opposite direction at the time of
the explosion are now 780 trillion miles from
this portion of the nebula. Due to the steady
decline in velocity caused by these collisions,
the nebula will cease to glow in another
25,000 years. The light captured by the
telescope to make this picture left the nebula
about 2,500 years ago. Photographed with the
48-inch Schmidt camera at Palomar*

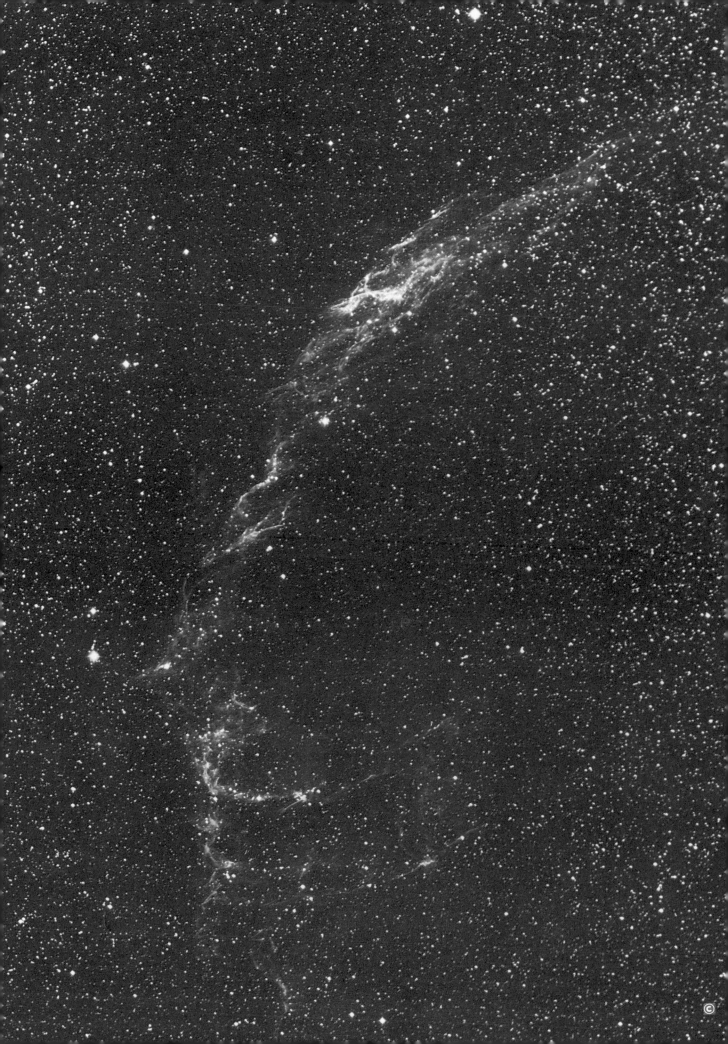

GALAXY IN SCULPTOR (upper), photographed with the 48-inch Schmidt reflector at Mount Wilson. SPIRAL GALAXY IN PEGASUS (lower), photographed with the 200-inch reflector at Mount Wilson

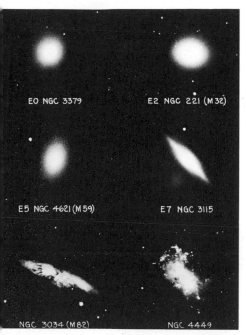

VARIOUS TYPES OF GALAXIES; *elliptical and irregular. Photographed with the 60-inch reflector at Mount Wilson*

VARIOUS TYPES OF GALAXIES: *spiral and barred spiral. Photographed with the 60-inch reflector at Mount Wilson*

them much more brilliant than the Sun. Such a conception would have been regarded as absurd even sixty years ago, but we know it to be true.

Our Galaxy appears to be a fairly typical system. The Andromeda Spiral is admittedly rather bigger, but it too contains all the familiar features, including globular clusters and even satellite galaxies which appear to be of much the same size as our Nubeculæ. This must also apply to many of the rest, even though only a few are close enough to be studied in any detail.

Many problems remain to be solved. For instance, we have as yet no idea why some of the galaxies are spiral, and only recently has it been proved that the rotation is with the spiral arms trailing. Neither do we know for certain whether an elliptical galaxy turns into a spiral, or vice versa; and we cannot even be sure whether all galaxies go through a spiral stage at some stage during their evolution. Most astronomers now think not.

It is significant that elliptical galaxies are mainly Population II. Most of the interstellar material has presumably been used up, and the hot early-type supergiants have disappeared, to be replaced by vast red stars which are more advanced in their careers. It is tempting to suggest, then, that spirals are at an earlier stage than ellipticals, but it is now thought unlikely that the different forms represent a genuine evolutionary sequence. It has been suggested that the key to the problem is the original rate of spin, so that quick-spinners will become spiral and slow-spinners elliptical. At the moment we have no definite answer, and our ideas are changing all the time.

Meanwhile, there is another point to be considered—the movements of the galaxies. Proper motions are too small to be measured, but radial velocities are not, and in 1920 V. M. Slipher, working at the observatory which had been founded in Arizona by Percival Lowell, made yet another remarkable discovery. Apart from the Andromeda Spiral and a few others which we now know to be members of our local group, all the galaxies showed Doppler shifts to the red end of the spectrum. This meant that all were receding from us.

The discovery was a complete puzzle at the time, because it was still not certain whether or not the galaxies were true external systems. When Hubble made his classic observations of Cepheids in the Andromeda Spiral, the situation became even stranger. If the red shifts were to be believed, all the galaxies beyond the local group were running away; and the more remote they were, the faster they seemed to go.

Using the 100-inch Hooker reflector, Hubble and his colleague Milton Humason probed to 700 million light-years, which was the distance of a cluster of galaxies in Ursa Major. The speed of recession turned out to be 26,000 miles per second. After the Second World War, the Palomar telescope came into use. With its superior light-grasp it could reach out still farther, and the red shifts showed that the faint galaxy 3C–295 is racing away at not much less than 90,000 miles per second.

It seems unbelievable. If you listen to a grandfather clock which ticks once per second, it is hard to realize that between each pair of ticks the Boötes galaxy recedes another 90,000 miles. Yet all the

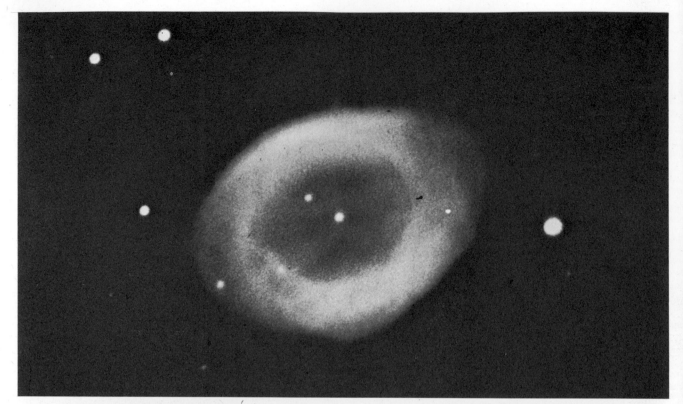

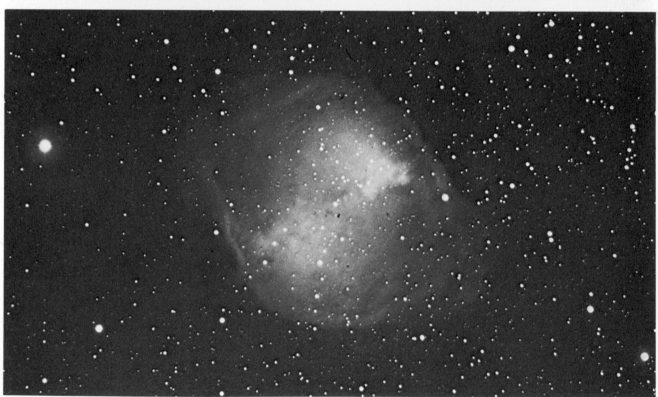

THE RING NEBULA IN LYRA, M57 (*upper*). *The famous planetary, photographed with the 200-inch Hale reflector.* THE DUMB-BELL NEBULA IN VULPECULA, M27 (*lower*). *Another planetary, much less symmetrical than the Ring. 200-inch photograph*

THE NORTH AMERICA NEBULA IN CYGNUS, NGC 7000 (*upper*). *Photograph taken with the 48-inch Schmidt at Palomar.* THE TRIFID NEBULA IN SAGITTARIUS, M20 (*lower*). *200-inch photograph. This is a gaseous nebula, in our own Galaxy*

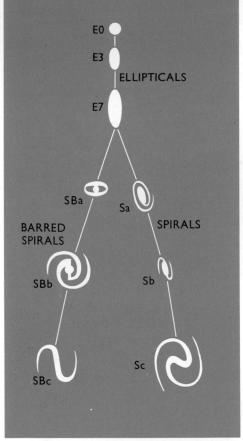

E0

E3

ELLIPTICALS

E7

SBa

Sa

BARRED
SPIRALS

SPIRALS

SBb

Sb

SBc

Sc

HUBBLE'S CLASSIFICATION OF THE
GALAXIES: *ellipticals (E0 to E7), spirals
(Sa, Sb, Sc), and barred spirals (SBa, SBb,
SBc)*

evidence points the same way. Unless there is a major mistake, and the red shifts are not Doppler effects at all, the estimated velocities must be at least of the right order.

There is no suggestion that our Galaxy is in any way a special case. The spectroscopic measures show that each cluster of galaxies is receding from each other cluster, so that the whole universe is expanding.

At 90,000 miles per second, the Boötes galaxy is receding from us at roughly half the speed of light; and if the majority view is accepted, some quasars are receding at over 90 per cent of the velocity of light. There must be a limit. If the recessional speeds continue to grow with increasing distance, there must come a stage when we consider objects which are moving away at the full 186,000 miles per second. No light could reach us from such an object; it would have passed over the boundary of the observable universe.

No optical telescope can probe far enough for us to decide whether or not the law known as Hubble's Law—the greater the distance from us, the greater the speed of recession—holds good out to the edge of the observable universe (if, indeed, there is such an edge). Yet once again we can call upon a modern branch of science which may be able to provide an answer. This is radio astronomy, which began only in the 1930s, but which has already more than proved its worth.

CLUSTER OF GALAXIES IN HYDRA, *photographed with the 200-inch Hale reflector at Palomar. Foreground stars of our own Galaxy are shown, but many of the blurred patches are due to remote external galaxies, each of which is composed of perhaps 100,000 million suns*

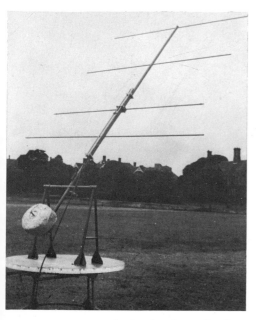

A YAGI ARRAY, *the simplest form of radio telescope*

THE ELECTROMAGNETIC SPECTRUM. *The central strip is the 'optical window', indicating visible light; it is clear that we can 'see' only a very small part of the total electromagnetic spectrum. The Earth's atmosphere will pass only visible light (the optical window) and the radiation in the 'radio window'. This is one reason why research with high-altitude intrumented rockets and space-probes is now so important*

THE COLOR OF LIGHT depends on its wavelength. As we have seen, red light has a longer wavelength than blue or violet. The unit of length is the Ångström, named in honor of the Swedish physicist Anders Ångström, who was born in 1814 and died in 1874, and who followed up the work of Kirchhoff, carrying out many valuable experiments. One Ångström unit is equal to one hundred-millionth of a centimetre. Red light has a wavelength of about 7600 Å; violet light, about 4000 Å.

Now suppose that we have radiation whose wavelength is longer than 7600 Å? It will not affect our eyes, and so we will not be able to see it, but it will be measurable in other ways. Most people are familiar with the lamps used in hospitals. The radiation emitted by these lamps is *infra-red*, and is simply too 'long' to be seen visually. Similarly, short waves below 4000 Å, the *ultra-violet*, are invisible to us.

It is rather surprising to find that the range of visible light, from 7600 to 4000 Å, is only a very small part of the *electromagnetic spectrum*. Beyond the ultra-violet, we have the invisible X-rays and gamma-rays; beyond the infra-red, we have radio waves.

To make matters worse, there are layers in the upper part of the Earth's atmosphere which block out most of the electromagnetic spectrum. Consequently much of the radiation coming from space does not reach us at all, but is stopped by the atmosphere. We have the so-called *optical window*, which includes the visible light from red to violet, and some distance into the infra-red and ultra-violet; we have the *radio window*, which allows some of the much longer waves to pass through. Otherwise, we can record nothing—so long as we stay on the surface of the Earth. To study the rest of the radiations, we must send our instruments up above the troublesome absorbing layers. Modern rockets are able to do this without the slightest difficulty, but the situation was very different in the year 1931, when a young research worker named

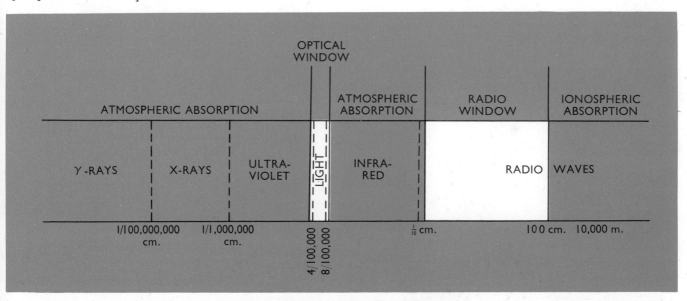

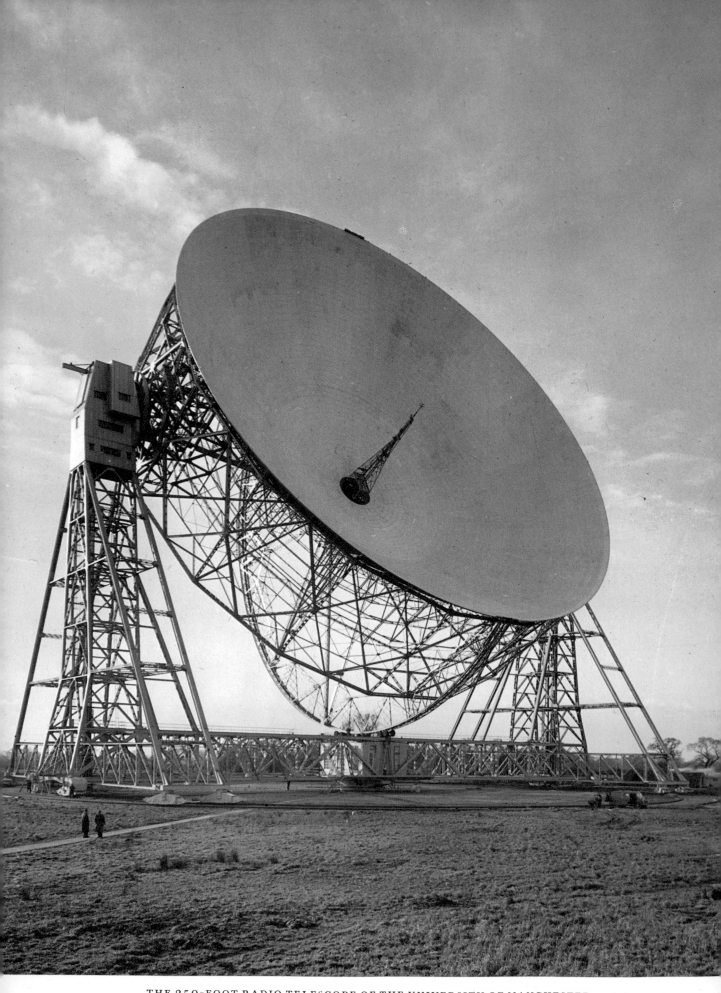

THE 250-FOOT RADIO TELESCOPE OF THE UNIVERSITY OF MANCHESTER,
*which was opened at Jodrell Bank, in Cheshire, in 1957. It is the largest 'dish'-type radio
telescope in the world, and is designed for automatic following of celestial objects across the
sky. It is made of steel, and weighs 2,000 tons*

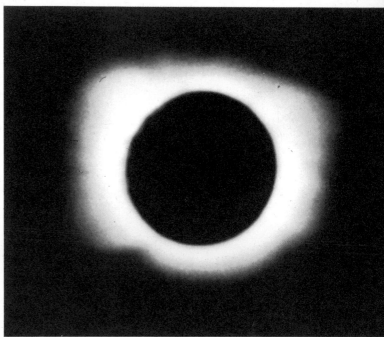

Above left

RADIO TELESCOPE AT ARCETRI.
One of the Italian radio telescopes, at Arcetri Observatory near Florence. It is entirely different in type from the 'dish' instrument such as the 250-foot paraboloid at Jodrell Bank. Photograph by Colin Ronan, 1961

Above right

TOTAL ECLIPSE OF THE SUN, *October 2, 1959, photographed from a McDonnell F.101B Voodoo aircraft at an altitude of 45,000 feet. Radio astronomers are particularly interested in solar eclipses*

RADIO AERIALS AT CAMBRIDGE,
photographed by Patrick Moore, 1971

Karl Jansky made a discovery which ranked among the most important of the twentieth century.

Karl Guthe Jansky was born in Oklahoma, in the United States, in 1905. His father was of Czech descent, but had settled permanently in America and had become a Professor in the University of Wisconsin. Karl Jansky took his degree in physics at the same University, and then joined the Bell Telephone Laboratories. His main work was to carry out research into problems of short-wave radio communication, and he was particularly concerned with 'static'. Static, known commonly though not scientifically as 'hissing and crackling', is the radio operator's worst enemy, and the Bell authorities wanted to find out as much about it as they could.

Jansky set up a strange contraption on a farm at Holmdel, New Jersey. It was an experimental radio aerial, looking somewhat like the skeleton of an aircraft wing, and it was driven round by means of a motor. The wheels were taken from a dismantled Ford car. Jansky's 'merry-go-round', as it was nicknamed, was designed to investigate static, but it achieved something much greater than anyone could have hoped.

There was plenty of static. There was radio noise due to near-by thunderstorm activity, and more noise due to more distant storms over 100 miles off. There was also a third noise; a very weak, steady hiss in the receiver. The source of this noise remained a puzzle for some time. It was not close at hand, and it seemed to come from a definite point in the sky. The source-point moved from day to day, and eventually Jansky found the answer. The hiss came from the Milky Way; more accurately, from that part of the Milky Way which lies in the constellation Sagittarius.

Let us go back for a moment to the shape of the Galaxy. We

KARL JANSKY'S 'MERRY-GO-ROUND', *the first true radio telescope. It was with this equipment that Jansky first detected radio emission from space, and so founded the science of radio astronomy*

know that the Sun is some 32,000 light-years from the centre, and that the nucleus of the Galaxy lies in the direction of the Sagittarius star-clouds. This, too, was the position of Jansky's radio source. Could it be that the long-wave radiation picked up by the 'merry-go-round' was coming from the very centre of the Galaxy?

Jansky believed so, and he published his results. Technical journals referred to them, and on May 5, 1933, when the news was released, some of the American daily papers carried headlines about it. It is very curious, then, that little attempt was made to follow the matter up. It is even stranger that Jansky himself did almost no further research in radio astronomy, the science which he had created. He published a few more results in 1937, but after that he abandoned the problem altogether. He died in 1949.

One point must be made clear at once. Sound-waves are carried by air. There is no air above a height of a couple of thousand miles above the Earth at most, and so no sound-waves can cross space. Jansky's radio hiss was produced inside the instrument itself, and was merely a means of recording the long-wavelength radiations. This also applies to modern radio telescopes of all kinds. We cannot 'hear' noise from space, but merely pick up the radiation and then convert it into a hiss. (In these days, better recording methods are available for most purposes.)

Professional astronomers of the time were uninterested, and Jansky's pioneer work was almost ignored. Fortunately it was followed up by a brilliant amateur, Grote Reber, who was by profession an electrical engineer. He had been born in Chicago in 1911, and even when still a boy had built powerful transmitters with which he communicated with other radio enthusiasts all over the world.

In 1937 Reber turned his attention to the problems of what he called 'cosmic static', and he built the first modern-type radio telescope. It consisted of a 'dish' or parabolic metal mirror, 31 feet 5 inches in diameter, operating on a wavelength of about 2 metres. The mirror focused radio waves just as the mirror of an ordinary telescope focuses light waves, but of course no visible image was produced. One does not 'look through' a radio telescope.

Reber confirmed Jansky's discovery of radio waves from the Sagittarius star-clouds. He also found other sources in Cygnus, Cassiopeia, Canis Major and Argo. What did surprise him was

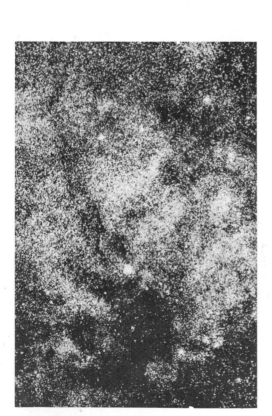

STAR-FIELDS IN THE MILKY WAY, *photographed at Mount Wilson*

I 7-FOOT 'DISH'-TYPE RADIO TELE-SCOPE IN AUSTRALIA. *It operates at a wavelength of 1 to 2 metres*

PULKOVO RADIO TELESCOPE. *One of the Russian radio telescopes at Pulkovo Observatory, Leningrad. Its main feature is a 'rail' aerial of 90 separate sheets; it has high resolution, and works at a wavelength of 2 to 5 c.m. Photograph by Patrick Moore, 1960*

that these sources did not agree with the positions of visible stars. For instance, no radio waves could be detected from Sirius, Vega or other brilliant objects; yet a part of the sky in Cassiopeia, unmarked by any bright star, seemed to send out strong radio waves.

Reber published some of his findings during the years between 1940 and 1945, when the war was going on. Later he set up some equipment on Haleakala, an extinct volcano in the Hawaiian Islands, to study radiation of even longer wavelength, and in 1954 he went to Australia to continue his researches. Modern radio telescopes make his original 31½-foot 'dish' look very small, but there can be no doubt that the greatest pioneer of radio astronomy was Grote Reber.

One of Reber's early ideas was to bounce energy waves off the Moon and record the 'echoes'. He was not successful at the time, but radar developments during the war paved the way for specta-cular advances in true radio astronomy.

If you go into a darkened room and shine a torch on to the far wall, you will see the wall, because it reflects the light back to your eye. There is some comparison—though not an accurate one—with the way in which a tennis-ball rebounds from a wall and so may be caught. If then we send out a radio 'pulse', it ought to be possible to record the 'echo' as the pulse hits a solid object and is reflected back to the transmitter.

This is the principle of *radar*, originally known as radiolocation. The method was developed during the Second World War, mainly by British scientists. It allowed the various ground forces to detect enemy aircraft, and to give the range and direction of attacking machines; it also led to various new instruments used in navigation.

To give further details about radar would be beyond the scope of the present book; suffice to say that it was discussed by R. A. Watson Watt as long ago as 1935, so that it is almost as old as radio astronomy itself. Meanwhile, radar echoes have indeed been received from various celestial bodies. The Moon echoes were first recorded in 1946 by the Hungarian scientist Z. Bay, and since then many experiments have been carried out in Britain, Australia, the United States and Russia. Radar echoes have also been received from Venus, Mars, Mercury and even the Sun. In addition, meteors may be studied by radar, since the trail left by a meteor acts, from the 'echo' point of view, rather in the manner of a solid body—though of course, it lasts for only a second or two. One interesting discovery has been that many meteor showers occur in the daytime. Such meteors cannot be seen (unless they are exceptionally brilliant), but radar is un-affected by the lightness of the sky and is equally unaffected by cloud. In consequence, radar has largely superseded the old methods of visual meteor-watching.

Radar development was in full swing during the year 1942, when the war was at its height. British anti-aircraft gunners were using radar to detect approaching German machines, but it was found that some mysterious 'jamming' was periodically upsetting all the equipment. It was suggested that the Nazis might have invented some means of putting the radar out of action, and a

THE JODRELL BANK RADIO TELESCOPE, *viewed from the control desk. As well as recording radio waves from space, the telescope has been used to track the various artificial satellites and space-probes launched since 1957*

AYSTACK, *the American radio telescope,*
photographed by Patrick Moore in 1964. The
equipment is inside the 'dome' which acts as
protective shield.

ARK II PARABOLOID AT JODRELL
ANK (below). *This is slightly different in*
appearance from the more familiar 250-foot.
Photograph by Patrick Moore, 1967. GREEN
ANK RADIO TELESCOPE (below
ght), *photographed by Patrick Moore in*
West Virginia

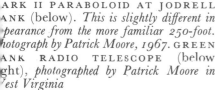

LONG-ENDURING METEOR TRAILS, *which act as reflectors of radar pulses. By radar studies, our information about meteors has been vastly extended during the last two decades*

research team under J. S. Hey undertook to find out what was happening.

It did not take Hey and his colleagues long to find out the source of the jamming. It did not come from a German transmitter; it came from the Sun.

Years earlier Karl Jansky had looked for radio noise from the Sun, and had failed to find it. This was not Jansky's fault. His work had been done near the time of a sunspot minimum, when the Sun was at its least active. In the early 1940's, when Hey investigated the matter again, there were more frequent large sunspots, and these produced bursts of radio noise. Most powerful of all were the flares, short-lived phenomena at a higher level than the spots.

When Hey first made his discovery, it came under the heading of a military secret. Other nations on both sides were busy working at radar development, and very little more research in pure radio astronomy could be done until the war was over. As soon as possible after the end of fighting, scientists turned back to the radio investigation of the universe. By now it was clear that the new methods could add tremendously to our knowledge, and the importance of the pioneer work of Jansky and Reber was at last realized.

An American, George Ellery Hale, was mainly responsible for the world's greatest optical telescopes, including the Palomar 200-inch. In radio astronomy, Hale's role has been played by a Briton—Professor Sir Bernard Lovell, one of the radar pioneers. Just as Hale planned instruments of greater size and power than any built in earlier years, so did Lovell. The aim this time was a giant radio telescope with a 'dish' 250 feet in diameter.

Lovell had to face immense difficulties. There were the usual money troubles; Britain, after all, had been financially crippled by the war. Moreover, radio astronomy was still a relatively new science, and the projected 250-feet telescope presented all manner of design problems. By his personal enthusiasm and technical skill, and helped by a brilliant research team including R. Hanbury Brown, another of the radar pioneers, Lovell overcame the difficulties one by one. The 250-feet 'dish' was built, and set up at the Jodrell Bank Research Station near Manchester.

Jodrell Bank has been to radio astronomy what Mount Wilson and Palomar have been to optical astronomy. Even to list the discoveries made there would take pages. The original 250-foot paraboloid has been modified, and remains the largest of its kind; it has now been joined by a 210-foot 'dish', and elaborate new projects are under way. Other instruments have been set up elsewhere in the world—for instance at Parkes, in Australia, where the 210-foot radio telescope can study southern objects inaccessible from England; and at Arecibo, in Puerto Rico, there is a 1,000-foot paraboloid, though it has been built into a hollow in the gound and is not steerable. Radio astronomy has become of vital importance in modern research; Lovell will always be honored as the man who played the key rôle.

The Jodrell Bank 'dish' focuses radio waves, just as an optical mirror focuses light waves. The focal point is the tip of a 60-foot-long metal rod or *dipole* which sticks out from the centre of the disk; the radio waves coming from space are reflected to this point, and are amplified. They are then passed to an automatic pen which records them on a moving chart paper.

Not all radio telescopes have 'dishes'. There is, for instance, the Mills Cross type, named in honor of the Australian scientist Bernard Y. Mills, who was one of the main designers (another was a Briton, Sir Martin Ryle). Here we have two rows of aerials set up perpendicular to each other. The first receives the radio waves from one area of the sky; the second receives the waves from another part. If both rows are receiving together, there will be double strength signals corresponding to the centre of the cross. If the rows are connected so as to work in opposition, the strength in the centre of the cross will be zero, since one row will cancel out the other. By means of a switch, the connections of the rows are rapidly changed from one position to the other, and the current in the centre of the cross will vary up and down when a radio source lies in the position defined by the centre of the cross. By this method the position of the radio source itself can be worked out.

Now let us see what sorts of objects send out radio waves.

The Sun is one source. Another, surprisingly, is the planet Jupiter; it is thought that the erratic radio emissions from it are linked with the Jovian magnetic field, which is probably very powerful. Yet most radio sources lie far beyond the Solar System. Pride of place must go to the Crab Nebula in Taurus, which has already been described, and which is the wreck of the 1054 supernova; other galactic sources are also thought to be supernova relics. This is probably true also for the strong source Cassiopeia A, which does not correspond in position with any bright star or recorded supernova. It seems to be about 10,000 light-years away, and the Palomar reflector has been able to photograph a very faint luminous gas-cloud in the exact position of the source.

Toward the end of 1967 Miss Jocelyn Bell (now Mrs. Burnell), one of the radio astronomers at Cambridge, discovered a strange, quickly-varying radio source with a period of only 1·3 seconds. This was the first pulsar. It caused intense excitement, and for a brief period it was then thought possible that the signals might be artificial—though this attractive idea was soon discounted!

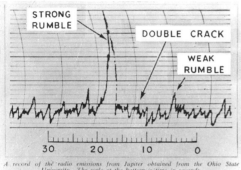

RECORD OF RADIO EMISSIONS FROM JUPITER. *The scale gives time-intervals in seconds*

MILLS CROSS: *looking along the north–south arm*

©

THE CRAB NEBULA, MESSIER I IN TAURUS. *Here we see the result of a stellar explosion which occurred over 6,000 years ago. Due to its distance from the Earth, the explosion was not observed until 6,000 years after it had taken place. On the morning of July 4, 1054, a star was seen by Chinese observers, shining with a brightness 100 million times greater than previously. After several months it faded below naked-eye visibility. In the place of this star or 'supernova', modern telescopes now reveal this large cloud of gas, still expanding at a rate which increases the diameter of the nebula by nearly 70,000,000 miles each day. High-energy electrons, still moving about rapidly as a result of the explosion, cause the centre of the nebula to glow with nearly white light, and at the same time cause the wispy filaments of gas to shine with characteristic color. In the gas-cloud is a pulsar or neutron star, which is almost certainly the remnant of the supernova itself, and may be termed the 'power-house' of the Crab. Photographed with the 200-inch Hale reflector at Palomar*

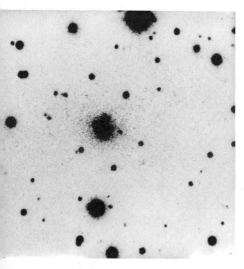

RADIO SOURCE CASSIOPEIA A, *a particularly intense source. (This is a negative photograph.) The emission is due to distributed masses of gas, and there is no bright visual object*

Many pulsars have now been detected, though only one, that of the Crab Nebula, has been identified with a visual object.

Pulsars were originally thought to be special kinds of White Dwarfs, but it is now believed that they are neutron stars, even smaller and denser. According to the Russian astronomer Vitaly Ginzburg, a pulsar has a solid outer crust which is not made up of neutron material, and which can experience 'star-quakes' resulting in the abrupt changes of period which are noted in some pulsars. It may well be that the pulsar is rotating, and that the pole of rotation is not coincident with the magnetic pole; as the object spins, we come regularly into the 'beam of emission', rather as a watcher on the coast passes regularly through the beam of a lighthouse. In any case, a pulsar may represent the very last stage in the life of a star. Many dead pulsars must exist in the Galaxy, which we cannot detect because they are sending out neither light nor radio waves.

Interstellar material also sends out radio emission. We know that there is a great deal of gas spread through the Galaxy. Of this gas, hydrogen is much the most plentiful, and tends to collect into huge clouds, each some 30 light-years across. It is of course extremely cold, and also extremely rarefied; there are less than ten atoms per cubic centimetre, which is a low density judged by any standards. Nevertheless, in 1944 a Dutch scientist, Hendrik van de Hulst, predicted that this hydrogen should be sending out radio waves on a length of 21·1 centimetres. He knew that it would be weak and hard to detect, but he believed that it must be present.

At the time, van de Hulst was unable to carry out experimental work. The Germans occupied Holland, and scientific research was at a standstill. It was not until 1951 that two Americans, H. Ewen and E. Purcell, managed to detect the 21·1-centimetre radiation; but when they did, it proved to be most important.

Interstellar gas is most plentiful in the spiral arms of any galaxy. Moreover, if it is in motion, it will yield Doppler shifts in the same way as visible light. By plotting the positions of the hydrogen clouds, and observing their Doppler shifts, radio astronomers were able to prove what had already been suspected—that our own Galaxy is spiral in form. By now five definite spiral arms have been traced, and in addition we can receive radio waves from the galactic nucleus—since these waves are not blocked by obscuring matter, but penetrate right through.

We can see the way in which the story has unfolded. Herschel gave a good idea of the disk-shape of the Galaxy; the distribution of bright stars of early spectral type indicated the possibility of a spiral form, and radio astronomy proved it. There is, after all, a strong link between Herschel and van de Hulst.

Beyond the Galaxy lie other radio sources. One of these is the Great Spiral in Andromeda. Reber, before the war, predicted that radio waves would one day be received from the Spiral, and his forecast has come true. But what of the remarkable source in the Swan, known as Cygnus A? Apart from Cassiopeia A, it is the most intense radio source in the sky—but its distance is much greater. One estimate places it at 700 million light-years.

Up to a few years ago, it was thought that Cygnus A and others

RADIO SOURCE CYGNUS A. *Another intense radio source. This is a negative photograph, obtained at Palomar*

N.G.C. 1275, ONE OF THE MEMBERS OF THE PERSEUS CLUSTER OF GALAXIES, *photographed with the 200-inch Hale reflector at Palomar. This galaxy is a radio source*

of its kind might be due to galaxies in collision. The idea was that if two galaxies passed through each other, rather in the manner of two orderly crowds of people moving through each other in opposite directions, the individual stars would seldom collide, but the constant collisions between particles of interstellar matter would produce the radio emission picked up on Earth. Unfortunately for this idea, it has since been found that the process could not produce nearly enough energy, and the attractive theory of colliding galaxies has had to be given up, so that at present we have to admit that we do not know why certain kinds of galaxies are so strong in the radio range. There are galaxies, too, which seem to have suffered immense explosions in the remote past; one of these is Messier 82, in the Great Bear, which is an energetic radio emitter, and which shows intricate hydrogen-gas structures moving at velocities of up to 600 miles per second outward from what seems to be the explosion-centre. From studies of the gas-motions it has been found that the outburst in Messier 82 happened $1\frac{1}{2}$ million years before our modern view of it. Since the distance of the galaxy is 10 million light-years, the actual explosion dates back about $11\frac{1}{2}$ million years.

Meanwhile, some new objects have been studied which may cause us to undertake a drastic revision of all our theories about the make-up of the observable universe. These objects are the quasars, alternatively known as quasi-stellar objects or (for short) Q.S.O.'s.

The story of quasars began when Cambridge radio astronomers plotted the positions of radio sources in the sky, and tried to identify them with objects seen visually. In one case, 3C-48 (the 48th object in the third Cambridge catalogue of radio sources), the position seemed to agree with a faint star, strangely blue in color and with a peculiar spectrum. Then, in 1963, M. Schmidt, at Palomar, began a study of a similar object, 3C-273. The position of the radio source was known very accurately, and Schmidt identified it with an object which looked at first like a star. When he examined the visual spectrum, he found that the innocent-looking object was certainly not a star. Its spectrum was entirely different, and there was an immense red shift. If this red shift were due to a Doppler effect, then the object must be immensely remote and super-luminous. Subsequently, other 'quasars' were found; the most remote of them appear to be receding from us at over 90 per cent of the velocity of light.

Quasars are, as yet, very much of an enigma. If the red shifts are taken at face value, then one quasar must be the equal of about 200 whole galaxies—and yet it is very small compared with a galaxy, so that it must be drawing upon some completely unknown source of energy. On the other hand, it is quite possible that there are serious mistakes in interpretation, and that the quasars are not nearly so remote or so luminous as their red shifts indicate. We can only await the results of further research.

Had it not been for radio astronomy, the curious nature of the quasars would still be unknown; they would still be regarded as ordinary faint stars. Moreover, radio astronomy can reach farther out than optical astronomy—and this brings us on to the most vital problem of all, that of the origin of the universe.

A QUASAR. *The quasar is shown by the arrows. It looks stellar, but is in fact immensely remote and luminous—assuming that modern theories are approximately correct*

In the 1930's, some pioneer work upon this problem was carried out by the Belgian abbé Georges Lemaître. Lemaître supposed that all the material in the universe was originally concentrated in one immensely dense 'primæval atom', and that between 20,000 and 60,000 million years ago this 'primæval atom' exploded, sending its material outward in all directions. For a time there was a balance between expansion and contraction, but then, about 9,000 million years ago, something or other tipped the scale in favour of expansion, so that the groups of galaxies are still receding from each other. On the strict evolutionary theory, the universe had a definite beginning, and will have a definite end; all the stars in all the galaxies will die, so that the universe may be compared with a clock which is running down.

The main theory has been modified since Lemaître's original papers were published, and, in particular, it is certain that there are errors in his time-scale; the expansion of the universe began over 9,000 million years ago. Also, it is important to note that neither Lemaître nor any of his successors has attempted to explain just how the original material came into existence—so that strictly speaking we are discussing not the origin of the universe, but its development.

After the war, a different scheme was proposed by H. Bondi and T. Gold, working at Cambridge. This 'steady-state' theory, popularized and modified by F. Hoyle, caused a great deal of discussion, and was certainly ingenious. It supposed that the universe has always existed, and will exist for ever. New matter is being created out of 'nothingness' all the time, so that as old galaxies die, new ones will take their place. Of course, the process will be very gradual. Matter is supposed to be created in the form of hydrogen atoms, and there is no chance of our being able to detect it, any more than one could track down the formation of a single new grain of sand in the Sahara Desert. Again there is no explanation of how matter could appear out of nothingness—but this weakness is also common to the evolutionary theory, so that on this score the one idea is as reasonable as the other. On the steady-state theory, the universe may be compared with a clock which is being continuously re-wound.

Yet when we study objects thousands of millions of light-years away, we are seeing them as they used to be thousands of millions of years ago; in effect, we are looking back into the past. Radio studies carried out at Cambridge by Sir Martin Ryle and his colleagues have shown that the universe in those remote times was not arranged just as it is today, and the steady-state theory has now been abandoned by almost all authorities. It is attractive, but it simply does not fit the facts.

This is not to say that the strict evolutionary theory is correct. It may well be that the universe is in an oscillating condition, so that the present phase of expansion will be followed by a period of contraction—and so on, perhaps indefinitely. As yet we cannot tell, but at least we have made definite progress in recent years, and research is continuing all the time. Radio astronomy holds out our main hopes, and we have certainly come a long way since that moment in 1931 when Karl Jansky used his 'merry-go-round' to detect the weak emission due to radio sources in the Milky Way.

N.G.C. 4038–9: RADIO GALAXIES, *photographed with the 48-inch Schmidt camera at Palomar*

THE SOUTH POLAR AREA OF MARS, 1888, *according to Schiaparelli; many canals are shown*

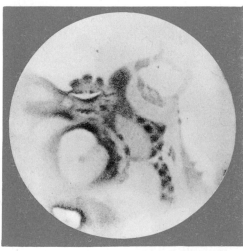

DRAWING OF MARS BY LOWELL. *Here we see the canal network in its most developed form*

MARS, 1909, DRAWN BY E. M. ANTONIADI. *The telescope used was the Meudon 33-inch refractor. The difference between Antoniadi's representation and those of Schiaparelli and Lowell is very obvious*

27 life in the universe

MODERN TELESCOPES, both optical and radio, can reach far out into space. We can photograph galaxies thousands of millions of light-years away, and we can measure the luminosities of stars whose light started on its journey towards the Earth long before life appeared on our world. Yet what of 'life' itself? Are we alone in the universe, or is our civilization only one of many?

Before we go any further, let us make sure what we know about 'life', so that we can narrow down our search as much as possible. We know that all the matter in the universe is made up of a comparatively few fundamental substances or *elements*; ninety-two are known to occur in nature.

Of all the various types of atoms, only one is able to produce the very complex atom-groups which are needed for living matter. This is the atom of carbon. All *organic* material, then, must be based on carbon. Many other elements are involved, but carbon is the fundamental basis. It follows that all life, wherever it may be found, is similarly dependent on the carbon atom. In this case it must be of the same kind as the life we know on Earth.

This may seem a sweeping statement—and indeed it is; it may well prove to be much too sweeping. First, it is worth noting that the silicon atom is a fairly good 'builder'. It is not nearly so efficient as carbon, and we have no evidence of the existence of silicon-based life; but according to some scientists, the possibility does exist.

There is another argument which is always raised in any discussion of life on other worlds. It runs as follows: 'Why should we say that all life in the universe must be built on the same pattern as ours? There may be intelligent races of quite different kind.'

The argument is plausible enough, and is impossible to disprove. Yet if we take all the evidence we have, and put the most reasonable interpretation on it, we have to agree that utterly alien creatures do not seem at all probable, either in our Solar System or beyond. If they exist, they must be built upon a pattern which we cannot understand, and so it is pointless to speculate. There is no proof that alien life-forms exist anywhere, and there is considerable indirect proof that they do not.

Of course, a being living on another planet need not be of human form. All we can say is that according to modern evidence, an extra-terrestrial being would be made up of familiar elements, and would be based—as we ourselves are—upon carbon.

Having made this point, we can start to look round for worlds which might be expected to support life.

There are two essential qualifications. A life-bearing world must be neither too hot nor too cold; intense heat breaks up organic molecules, and extreme cold prevents vital activity. It is also necessary to have an atmosphere, preferably one which contains a reasonable amount of oxygen and water-vapor.

Obviously we can rule out the stars, which have surface temperatures of several thousands of degrees. Of the nine planets in the

MARINER IX, *the Mars probe launched in May 1971, and which reached the neighborhood of Mars in the following November*

VALLEY IN RASENA AREA OF MARS, *photographed from Mariner IX in 1972*

Solar System, five (the giants, and Pluto) are too cold, while Mercury has no atmosphere. The asteroids are airless, and so are the satellites—except for Titan and possibly some of the major satellites of Jupiter, which must be rejected because of their low temperature.

Apart from the Earth, then, only Venus and Mars seem to be worth considering seriously. Before the Space Age the prospects did not seem too bleak. Venus was thought to have broad oceans, in which primitive life at least might exist, while the presence of vegetation on Mars was accepted by almost everybody—even though Lowell's canal-builders had long since been discredited. Unfortunately, what we have learned from the space-probes has caused a revision in outlook. Venus, with its temperature of about 900 degrees Fahrenheit and its very dense carbon-dioxide atmosphere, must be ruled out. We cannot yet prove that Mars, too, is lifeless; but the evidence points that way, particularly if the polar caps are made up of solid carbon dioxide rather than ice or snow.

There is always the chance that Earth-type organisms of lowly type might be persuaded to adapt to conditions on Mars, but the prospects are not good, and life for astronauts who go to the planet will have to be under highly artificial conditions. No: Mars, we must concede, has disappointed us. Certainly it cannot support life of advanced type, and in the Solar System we seem to be alone. Yet can there be planets circling other stars?

There have been many theories about the origin of the Solar System. One, the Nebular Hypothesis, was put forward in 1796 by Pierre Simon Laplace, a leading French mathematician. According to Laplace the system began as a vast gas-cloud, disk-shaped and in rotation. As the cloud cooled down, it shrank; and as it shrank its rate of spin increased, until the centrifugal force at the edge became equal to the gravitational pull there. At this stage a ring of matter broke away from the main mass, and gradually this ring condensed into a planet. As the shrinkage continued, a second ring broke away, to produce another planet; and so on. The final result was a central Sun surrounded by a family of smaller bodies.

The theory seems reasonable enough, but unfortunately there are strong objections to it. For one thing, a gaseous ring formed in such a way could not possibly condense into a planet.

In 1900 two American scientists, Chamberlin and Moulton, put forward another theory, according to which the planet-forming material was pulled off the Sun by the action of a second star which passed near by. This 'tidal' theory was modified by Sir James Jeans, who will be well remembered by many people not only for his astrophysical researches but also for his popular books and broadcasts.

This time we have an encounter between the Sun and a passing star which resulted in a 'tongue' of matter being pulled away from the Sun. After the second star had gone on its way, the 'tongue' broke up into pieces, each of which became a planet. An attractive feature of Jeans' idea is that the 'tongue' would have been thickest in its middle parts, which is where we find the largest planets, Jupiter and Saturn.

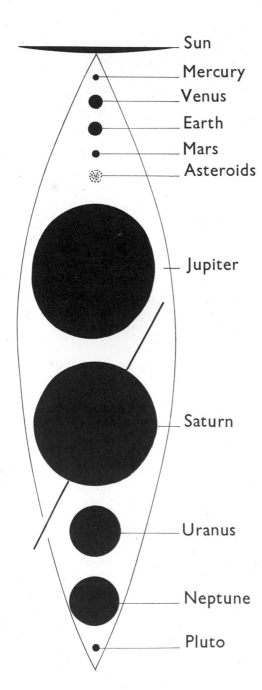

Sun
Mercury
Venus
Earth
Mars
Asteroids

Jupiter

Saturn

Uranus

Neptune

Pluto

JEANS' TIDAL THEORY OF THE
ORIGIN OF THE SOLAR SYSTEM. *The
theory was widely accepted for some years, but
mathematical investigations have revealed so
many weaknesses in it that it has now been
rejected*

Here again there are mathematical difficulties. Although the tidal theory still has its supporters, most scientists agree that it cannot be correct.

F. Hoyle once suggested that the Sun formerly had a binary companion, which turned into a supernova. Hoyle went on to calculate that the effects of the explosion would give what was left of the star a recoil sufficient to break it free from the gravitational pull of the Sun. He suggested that just before its final departure, it ejected a cloud of gas which the Sun managed to retain, and that the planets then condensed out of this material.

The most favored theory at present is based on work by a German, Carl von Weizsäcker, and a Russian, Otto Schmidt. This time it is supposed that some thousands of millions of years ago the Sun passed through an interstellar cloud, and collected an envelope of matter out of which the planets and other bodies of the Sun's family were formed.

It is too early to say whether this last theory is correct, but at least it seems to be promising. If we accept it, planetary systems must be common in space. What can happen to the Sun can happen to other stars as well.

The trouble, of course, is that a planet has no light of its own, and even a large body accompanying a relatively close star would be too faint to be seen even from Palomar. The only way in which such a body could be detected would be by its gravitational pull on the parent star.

In 1942, K. A. Strand published results indicating that the fainter member of the 61 Cygni binary system is associated with an invisible body with about 15 times the mass of Jupiter. However, the most convincing case is that of Barnard's Star, the faint Red Dwarf at a mere 6 light-years from us. The proper motion has been studied with great care by P. van de Kamp at the Sproule Observatory in America, and it now seems certain that the star is attended by at least one planet. More probably there are two. Van de Kamp believes that both are comparable in mass with Jupiter, and that their distances from the parent star are respectively 260 and 440 million miles, giving periods of 12 years for the inner body and 26 years for the outer. Since they will be of about the 30th magnitude, they are much too faint to be seen or photographed even with the world's largest telescopes. Were they of mass comparable with the Earth, they would produce no observable perturbations in the motion of the star, and we could know nothing about them.

Of course, all we can really say is that the van de Kamp objects are non-luminous bodies. We have no absolute guarantee that they are planets; but it is hard to imagine what else they could be. There is no question of their being lightweight stars.

If we have evidence of several planetary systems in our immediate part of the Galaxy, it follows, logically, that such systems are common. Indeed, it would be unreasonable to believe otherwise. Our Sun is one of 100 thousand million stars in the Galaxy; the Palomar reflector is capable of photographing 1,000 million galaxies. Therefore, the number of stars which we know to exist is 100 thousand million multiplied by 1,000 million —and even this represents only one part of the universe. It is absurd to suggest that in all this host, the Sun is unique in possessing a system of planets.

If the number of planet-families runs into hundreds of millions, some of these worlds must be suitable to support life. Where the conditions are right, life may be expected to appear. At this moment, some astronomer living on a planet circling a far-away star may be turning his telescope towards the yellow star we call the Sun, and wondering whether it also may be the centre of a 'Solar System'.

Interplanetary travel is a real possibility, but reaching the stars is another matter altogether, and may always remain beyond our powers. The only way of contacting distant races would be by means of radio signals, which travel at 186,000 miles per second—and in 1960 some interesting experiments were begun by an American scientist, Frank Drake, using the 85-foot radio telescope at Green Bank, in Virginia.

Drake surveyed all the comparatively close stars in an attempt to decide which of them might be attended by Earthlike planets. Very feeble stars such as the Red Dwarfs held out little hope; it was better to search for stars which are not too unlike the Sun. Two in particular caught his attention: Tau Ceti and Epsilon Eridani. Both are of magnitude $3\frac{1}{2}$, so that they are clearly visible to the naked eye; both are of spectrum type Ko, and both are about 11 light-years away. Tau Ceti is nearly half as luminous as the Sun, while Epsilon Eridani has about one-third of the solar luminosity.

If these stars were accompanied by inhabited planets, and civilization had reached a level equal to our own, radio astronomy would have developed. This means that the 21·1-centimetre radiation would come under study, since this is the radiation emitted by the cold hydrogen clouds in the Galaxy. Drake's scheme was to 'listen out' at this wavelength in an attempt to detect any signals which came in a definite pattern, and which could not therefore be of natural origin. Another idea was to send signals at the same frequency, arranged in a regular, obviously artificial manner, so that the hypothetical scientists of the other planets could receive them and—perhaps—reply.

The idea was not absurd, but it was clear from the outset that the chances of success were remarkably slight. For one thing, a radio wave would take 11 years to reach either Tau Ceti or Epsilon Eridani, and a further 11 years would elapse before a return signal could be received, so that a message transmitted in 1972 could not bear fruit until 1994 at the earliest. In any case, there is not the slightest proof that either star is attended by a planet, inhabited or otherwise. Reluctantly, it must be said that the experiment was the longest of 'long shots', and it has now been discontinued.

Even so, there is no reason to doubt that intelligent life is widely scattered through the universe. The Sun is an ordinary star; the Earth is an ordinary planet; and neither is there anything remarkable about mankind. We may indeed be very low in the cosmical intelligence scale, but it is a measure of our scientific advance during the last few centuries that we have come to realize how unimportant we really are.

GROUP OF GALAXIES IN LEO, *photographed with the 200-inch Hale reflector at Palomar*

THE 85-FOOT RADIO TELESCOPE AT GREEN BANK, VIRGINIA, *operated by the Associated Universities, Inc., under contract with the National Science Foundation. National Radio Astronomy Observatory, Green Bank, West Virginia*

28   rockets into space

EXOSPHERE
250 MILES
UPWARDS

IONOSPHERE

200
190
180
170
160
150
140
130
120
110
100
90
80
70
60
50
40
30
20
10
0

AURORA
POLARIS
(40-600 MILES)

METEOR
BURN-OUT

NOCTILUCENT
CLOUDS

HIGH TEMPERATURE BELT

STRATOSPHERE

NACREOUS CLOUDS

CIRRUS CLOUDS

CUMULUS   TROPOSPHERE
CLOUDS

SEA-LEVEL

CROSS-SECTION OF THE EARTH'S
ATMOSPHERE

THE SPACE AGE BEGAN on October 4, 1957, with the launching of the first artificial satellite: the Soviet vehicle Sputnik 1. Progress since then has been amazing by any standards. Less than eleven years separated Sputnik 1 from the first landing on the Moon.

In giving a very abbreviated story of rocket development, I realize that I am bound to overlap what has been said earlier in the present book; but there is no alternative, because the idea of space-flight is much older than many people imagine.

As long ago as the second century A.D., a Greek writer, Lucian of Samosata, wrote a story in which his heroes were sent to the Moon by the action of a waterspout, which caught the ship in which they were sailing and hurled it upward. Centuries later, a story by no less a man than Johannes Kepler described a lunar voyage. At about the same time an English bishop, Godwin, published an entertaining fantasy in which the journey to the Moon was made on a raft towed by wild swans. Of course, these and other tales were what we now call 'science fiction', and although some of them (notably Kepler's *Somnium*) contained some interesting scientific forecasts, the suggested methods of space-travel were—to put it mildly—impracticable.

In 1865 a better idea was put forward by Jules Verne, the great French story-teller. In his book *From the Earth to the Moon*, Verne planned to send his voyagers on their way by means of a space-gun. The projectile containing the travellers was put into the barrel of a vast cannon, and fired moonward at a speed of 7 miles per second.

Verne always kept to the proper facts as much as he could, and many of the calculations in his story were correct. For instance, he selected the right starting velocity. Seven miles per second is enough to enable a projectile to break free of the Earth's gravitational pull. It corresponds to about 25,000 m.p.h., and is termed *escape velocity*. Unfortunately there were two things which Verne failed to take into account. If a solid body were made to move through the dense lower atmosphere at such a speed, it would be so violently heated by friction that it would be destroyed at once. Moreover, the shock of starting at seven miles per second would certainly kill the luckless pioneers. No human frame could endure so tremendous a jolt.

Space-guns, then, are out of the question. Neither can we travel to the Moon by 'conventional' machines such as aircraft. An aircraft cannot function unless there is air round it, and above a height of a few miles there is so little atmosphere that for most purposes it may be disregarded. Most of the quarter-million mile flight to the Moon must be done in airless space. (Modern jet-aircraft are similarly limited, since their engines depend on being able to draw in oxygen from the surrounding atmosphere.)

Around the turn of the last century two men independently realized that the only way to reach the Moon was to use vehicles propelled according to what Sir Isaac Newton had termed the *principle of reaction.* In Germany, Hermann Ganswindt designed an extraordinary machine propelled by dynamite; in Russia, Konstantin Eduardovich Tsiolkovskii published the first papers outlining a basis for future research.

Ganswindt must have been a good approximation to the traditional inventor of the story-books. Most of his ideas were wild in the extreme, and few of his inventions really worked. For instance, he designed a helicopter, driven by a falling weight; but as he could not provide it with an engine, it was hardly likely to prove successful. He also described a new type of fire-engine, and even an airship. His 'space-ship' was bell-shaped, and packed with heavy steel cartridges filled with dynamite. Ganswindt's plan was to fire the explosive so that the upper part of the cartridge would bang against the top of the vehicle, thus knocking it upward. . . . A succession of such charges would, he reasoned, send the projectile into space, though in a rather jerky fashion.

Ganswindt's space-ship is as far-fetched as Lucian's waterspout launcher, and much less reasonable than Jules Verne's space-gun. Yet the germ of an idea was there, and it was unfortunate that Ganswindt had neither the patience nor the clear-headedness to follow it up. He died in 1934.

Tsiolkovskii was a very different sort of man. Unlike Ganswindt, he put forward his proposals in a truly scientific manner, and many of his theories still hold good today.

PREPARATIONS FOR FIRING VAN-GUARD *(left)*

PIONEER V LAUNCHER *(right). A 4-stage Thor-Able rocket blasting off from Cape Canaveral (now Cape Kennedy), Florida, carrying Pioneer V in its nose. The 94.8-lb. sphere, packed with scientific instruments, was launched into a 311-day orbit around the Sun between the orbits of the Earth and Venus. This was the third of the artificial planets; the first two were the American vehicle Pioneer IV, and the Soviet Lunik I*

LAUNCHING OF MARINER IX, *May, 1971. The probe begins its long flight to Mars*

LAUNCHING OF APOLLO 15, *in July, 1971, from Kennedy Space Center, Florida.*
Smoke billows as the huge Saturn V booster lifts off the pad

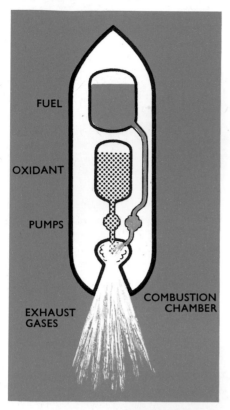

PRINCIPLE OF THE ROCKET

GODDARD'S ROCKET LAUNCHER AT
ROSWELL, 1934. *Goddard, the first man
to fire a liquid-propelled rocket, was one of
the greatest of the pioneers in this field*

He was the son of a forester, and was born at Izhevsk, in the
Russian province of Ryazanskii, in 1857. When he was a boy of
ten he caught scarlet fever, and was left permanently deaf, which
handicapped him all through his life. However, he was a skilful
mathematician, and became a teacher, first at a small country
school and then in the town of Kaluga.

Tsiolkovskii's first published article on the possibilities of space
travel appeared in 1903 in a monthly journal called *Na-utschnoje
Obosrenije* (Science Survey). It caused almost no interest even in
Russia, and nobody outside that country heard about it until
years later. Yet it was of great importance, since it laid down
many of the principles of the science we now term *astronautics*.

What Tsiolkovskii grasped was the vital fact that a rocket does
not need to be surrounded by air. The diagram will make this
clear. The inside of a firework rocket, for instance, is packed with
gunpowder; when the gunpowder is lit, hot gas is produced, and
this hot gas rushes out through the exhaust, so propelling the rocket-
body in the opposite direction.

Tsiolkovskii realized that gunpowder rockets of this kind are inefficient and unreliable. Instead, he planned to use liquid propellants. The two liquids, a fuel (such as petrol) and an oxidant
(such as liquid oxygen) would be pumped separately into a com-
bustion chamber; ignition would take place, and gas would be sent
out of the exhaust as with the firework. Using a rocket motor of
this kind would, he wrote, open up endless possibilities.

He was certainly right. Moreover, his claim to fame does not
rest only upon his two fundamental discoveries. Later he published
many articles, containing ideas which seemed absurd at the time
but which seem very far from absurd now. Of course he made
mistakes, but generally speaking he was well ahead of his time.

Tsiolkovskii retired from teaching in 1920. Four years later,
when the idea of space-flight had really started to take hold of
people's imaginations, the Soviet Government reprinted his
original article, following it with new productions of many of his
other works. Suddenly the shy, deaf ex-teacher became almost a
national hero. His seventy-fifth birthday, in 1932—three years
before his death—was marked by public celebrations, and Stalin
sent him a telegram of congratulation.

Tsiolkovskii was not a practical experimenter. He never fired a
rocket in his life, and probably never seriously considered doing
so. Yet nobody is likely to dispute his right to the title of 'father of
astronautics'.

Some years after the publication of Tsiolkovskii's first article,
an American scientist, Robert Hutchings Goddard, began to turn
his attention to rocketry. He had been born in 1882, at Worcester
in Massachusetts, and had become Professor of Physics at Clark
University. In 1914 he started serious experimental work, but the
war intervened, and not until after the Armistice was he able to
continue with his own research. Then, in 1919, he published a
monograph entitled *A Method of Reaching Extreme Altitudes*, in
which he stressed the value of rockets for upper-atmosphere
studies, and also suggested that it might be possible to send a
vehicle to the Moon. He did not discuss a manned rocket; he was
concerned with something much less ambitious—a small vehicle

carrying a charge of flash powder which would cause a visible spark as it hit the lunar surface.

The Moon-rocket proposal caused a tremendous amount of interest, and was featured in various American daily papers. Goddard was not pleased. He hated publicity in any form, and in any case he thought—rightly— that the Press was paying too much attention to the 'lunar shot' idea and too little to the immediate task of exploring the upper atmosphere of our own world. After that, he did his best to keep himself out of the public eye.

On March 16, 1926 he actually fired the first liquid-propellant rocket in history. It was powered by petrol and liquid oxygen, and rose modestly to a height of 200 feet, attaining a maximum speed of 60 m.p.h. It caused no comment at the time simply because almost nobody knew about it. Goddard published a report four years later, but it was not until 1936 that the date and nature of his pioneer experiment became generally known.

Meanwhile a remarkable book had been produced by a Rumanian teacher, Hermann Oberth. It appeared in 1923 under the title of *Rakete zu den Planetenräumen* (The Rocket into Interplanetary Space). It was mainly mathematical, but it quickly became something of a best-seller, because Oberth treated the whole space-travel problem in a properly scientific way.

PHOTOGRAPH OF THE EARTH, *as seen by the Apollo 8 astronauts, during their orbital flight around the Moon*

TOTAL ECLIPSE OF THE SUN, *February 15, 1961, photographed from Florence by Colin Ronan. The corona, shown here, is properly seen from the Earth's surface only during totality; full investigations of the Sun's surroundings depend on future research with rocket vehicles*

PHOTOGRAPHS OF THE AURORA BOREALIS (*Northern Lights*)

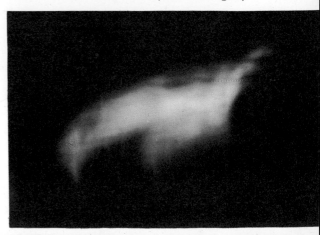

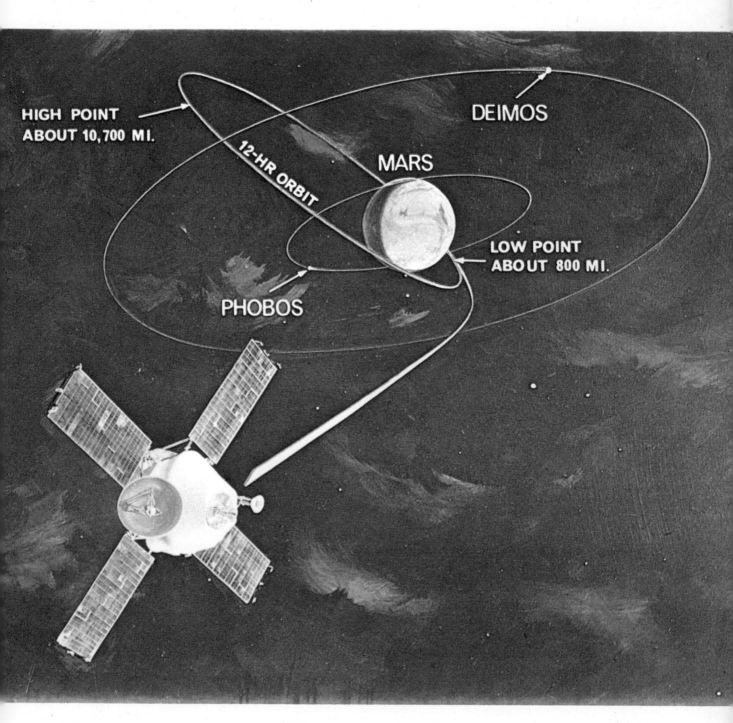

HIGH POINT
ABOUT 10,700 MI.

12-HR ORBIT

DEIMOS

MARS

LOW POINT
ABOUT 800 MI.

PHOBOS

ORBIT OF MARINER IX, November, 1971.
The diagram shows Mariner in orbit around
Mars, and also shows the orbits of the
satellites, Phobos and Deimos.

(Incidentally, his work was independent of Tsiolkovskii's. He could not read Russian—and still cannot, as he told me recently—and at that time Tsiolkovskii's works had not been translated into any other language.)

Oberth's book led to a surge of interest in astronautics, and Interplanetary Societies were formed in various countries. Particularly important was the German 'Verein für Raumschiffahrt', or Society for Space-Travel. Its leading members, such as Wernher von Braun and Willy Ley, decided to fire liquid-fuel rockets of

NIGHT LAUNCHING OF VIKING 13, *in the United States*

WOOMERA. *This is the main Commonwealth rocket range, and is situated in Australia. The photograph gives a general view*

their own, and actually did so. They knew nothing about Goddard's experiment of 1926, and naturally assumed that their vehicles were first in the field.

When the Nazi Party came to power in Germany, the military authorities realized that there might be a chance of turning the new rockets into weapons of war. The Verein für Raumschiffahrt ceased to exist. Some of its leaders, including Ley, left Germany; others, such as Von Braun, joined the official rocket station which was set up at Peenemünde, a small village on the Baltic island of Usedom. Work was continued, but this time for destruction instead of scientific progress. In 1937 came the first successful launchings, and on October 3, 1942, a much larger and more efficient vehicle, later known as the V2, soared high into the atmosphere.

As the war turned against Germany, the Nazis sent their V2 rockets against Britain. Between September 1944 and March 1945 vehicles carrying explosive charges fell on London and the Home Counties, and did considerable damage. The attacks ended when the launching sites were captured by the advancing Allied armies. Some of the German scientists, notably Von Braun, went to America, and the remaining V2s were shipped across the Atlantic to be used for scientific purposes.

Though the V2 was designed as a military weapon, it may be regarded as the ancestor of the space-ship. It could reach a height of well over 100 miles, which is beyond most of the Earth's atmosphere, and it was easy to adapt for high-altitude research.

There are many reasons for wanting to explore the upper part of the air. Until we can do so, we cannot hope to have a full knowledge of the atmospheric mantle. Moreover, remember that there are various absorbing layers which block out many of the radiations emitted by the Sun and stars. If we want to examine these radiations, we must send our instruments above the absorbing layers, and rockets give us our only means of doing so.

In addition there are the so-called cosmic rays—high-velocity particles, mainly atomic nuclei, which come from the depths of space, and which are bombarding the Earth all the time. When a cosmic ray *primary* smashes into the upper air it breaks up some of

SKYLARK ROCKET. *One of the British rockets being launched from the Commonwealth testing ground at Woomera, Australia. These vehicles were not designed to launch satellites, but to carry out upper-atmosphere research*

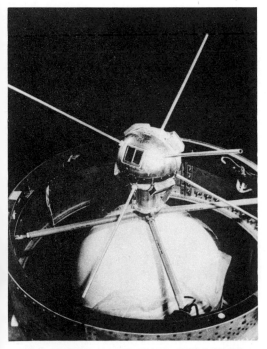

THE FIRST AMERICAN EXPLORER SATELLITE. *Though much smaller than the Russian vehicles, early American satellites carried complex instrumentation, and provided a great deal of valuable scientific information*

the air-particles, and is itself broken up. A whole series of collisions takes place, and only the broken pieces of the primaries reach the ground. This is fortunate for us, since the primaries may well be dangerous to living matter; but cosmic rays are clearly of fundamental importance, and scientists are anxious to study them as closely as possible.

As soon as the war was over the Americans began a serious programme of rocket research. The German V2s were improved, and new rockets soared high into the atmosphere—Aerobees, Vikings and the rest, each more powerful and more efficient than its predecessor. In 1949 came the first *step-rocket*, a development which requires a little explanation.

Even our best liquid propellants are not powerful enough to send a single-stage rocket beyond the atmosphere. Unlike the projectile of Verne's fictional space-gun, a rocket does not start off at full escape velocity, since if it were to do so it would be destroyed by frictional heating. It has to work up to maximum velocity more gradually, so that it travels through the dense lower air at a comparatively slow speed.

A step-vehicle consists of several rockets mounted one on top of the other. At the start of the flight, the bottom rocket does all the 'lifting'. When it has used up all its fuel, it breaks away and falls back to the ground; the second 'step' continues the journey under the power of its own motors, so that it has the double advantage of being well above the densest part of the atmosphere, and also moving at considerable velocity before its motors come into play. The second stage in turn can carry a third, and so on, though the design difficulties increase with each new step and there is a practical limit to the number of separate rockets.

The 1949 vehicle fired from the American proving ground at White Sands, New Mexico, consisted of a German V2 carrying a smaller rocket known as a WAC Corporal. Unaided, the WAC Corporal would have been able to reach fifty miles. Launched by the V2, it soared to almost 250 miles, which was a record at that time.

High-altitude rockets provided a tremendous amount of valuable information, but in some ways they were not satisfactory. No rocket can stay up for more than a few minutes. When it has reached the peak of its flight it plunges back to earth, and lands with a force sufficient to destroy it completely. Often enough the instruments are destroyed too, even if they are broken free during the descent and brought down by parachute. On July 29, 1955, the Americans therefore announced that they planned to send up a man-made moon or *artificial satellite*.

This was indeed a new departure. It meant that for the first time, men were planning to send proper scientific vehicles to carry out a research programme from above the Earth's atmosphere; it also meant that the true space age was about to open.

SPUTNIK I. *The first artificial satellite, launched by Soviet scientists on October 4, 1957*

TRAIL OF SPUTNIK II. *Dunfermline Abbey is shown in this photograph by Morris Allan*

29 satellites and space-flight

THE IDEA of an artificial Earth satellite was not new. It had been discussed by most of the rocket pioneers, including Goddard and Oberth. Yet the American announcement of 1955 came as something of a surprise, and the progress of what became known as 'Project Vanguard' was followed with intense interest.

Basically, the satellite, a small object containing scientific equipment, was to be taken up by rocket and put into an orbit round the Earth. Once it had been set into the correct path, it would not fall down, any more than the real Moon falls upon the Earth; the satellite would have become an astronomical body, and would behave in the conventional way. Of course, it would have to keep well above the main part of the Earth's atmosphere. If it entered the denser layers during any part of its orbit, it would be affected by air resistance, and would eventually spiral down to destruction in exactly the same manner as a meteor. When out in space, it would appear as a slowly moving star, crawling perceptibly against the background of the true stars.

Progress reports were made regularly from the United States base, then known as Cape Canaveral though since re-christened Cape Kennedy. Yet when success came, it was from an unexpected quarter—unexpected, that is to say, in Western Europe and the United States. On October 4, 1957, Russian scientists launched the first of all artificial satellites, known generally as Sputnik I. It took the form of a sphere 23 inches in diameter, weighing 184 pounds, and it was sent up in a multi-stage rocket. It carried a radio transmitter, and almost at once its 'Bleep! bleep!' signals were picked up by wireless operators in many countries. Its orbit, naturally enough, was elliptical; at its farthest point or *apogee* it was 588 miles above the ground, at nearest point or *perigee* only 142 miles. The time taken for it to complete one journey around the Earth was 96 minutes.

At less than 150 miles, the atmospheric drag is appreciable, and eventually Sputnik I came back into the air, destroying itself in meteor-fashion; but it lasted until January 4, 1958, by which time it had done all and more than its makers had hoped. By then the Russians had sent up a more massive satellite, Sputnik II of November 3, 1957, which took the form of a cylinder 70 or 80 feet long, and which carried complex instruments designed to study the upper air as well as phenomena such as cosmic radiation and solar ultra-violet. Sputnik II remained aloft until April 14, 1958; it was easily visible to the naked eye.

Meanwhile, the Americans had been having considerable trouble. The first attempt at a Vanguard launching, on December 6, 1957, resulted in an explosion which completely destroyed the rocket. Then, on January 31, 1958, the team led by Wernher von Braun successfully launched a tiny satellite, Explorer I, which weighed only 31 pounds, but which proved to be of great importance, since instruments carried upon it discovered the intense radiation-zones now called the Van Allen Zones.

MAJOR YURI GAGARIN. *The first space-traveller, Major Yuri Gagarin, made his pioneer flight on April 12, 1961 and orbited the Earth in the vehicle* Vostok, *landing safely in the prearranged position. He suffered no ill-effects from his flight, and the whole experiment was a complete success*

It seems a long time now since the launching of a new artificial satellite was given prominence in daily newspapers! Hundreds have been sent (400 by the Russian series of Cosmos vehicles alone). Most have been American or Russian, but smaller vehicles have also been launched by France, Japan, Italy and China. They have had many uses; we know far more about conditions above the atmosphere now than we did before 1957, and the practical applications in everyday life need not be stressed. Communications satellites alone have caused a major revolution in outlook. It no longer seems strange to look at a television screen and see what is actually happening on the other side of the world. Many lives have been saved; meteorological survey satellites have often given advance warnings of the development of dangerous storms. Science has benefited immeasurably from the coming of the man-made moons.

Manned space-flight began on April 12 1961, with the single-orbit journey of Yuri Gagarin in Vostok 1. The great break-through had been made; and after a second Russian flight, by Herman Titov, Commander Alan Shepard became the first American in space by making a sub-orbital 'hop' in May 1961. At that time the procedure was admittedly risky. Yet less than ten years later, Alan Shepard went to the Moon.

I have already said something about the lunar probes, automatic and manned, as well as the vehicles sent out to Mars and Venus. All these developments were crammed into the decade of the 1960s, which will surely be remembered long after the events of 1066 have been forgotten! Now, in the early 1970s, can we decide what will happen next?

One idea, warmly supported so long ago by Tsiolkovskii, is that of the fully-manned space-station. Certainly an orbital base would be tremendously useful, not only as an astronomical observatory and scientific laboratory but also as an intermediate station for space-craft bound for the Moon and planets. Elaborate designs were worked out, even before the war, by men such as Wernher von Braun; most of these designs envisaged a wheel-

APOLLO 15. *The astronauts and their LRV are on the plain at the foothills of the Apennines, with Hadley Delta, the 'uniform mountain', in the background*

shaped station, with the crew living in the rim and being provided with 'artificial gravity' in the guise of centrifugal force as the wheel rotated. Then, in the 1950s and part of the 1960s, the whole concept fell into disfavor. Now, with the American plans for a 'space-shuttle' and the Russian interest in orbital rendezvous maneuvers, it has been revived. It may well have been established before 1980; and from it, astronomers will be able to make observations which cannot be attempted from ground level. Fields of research such as infra-red astronomy and X-ray astronomy will be thoroughly opened up. Even before this, telescopes carried in automatic O.A.O. (Orbiting Astronomical Observatory) satellites will have been sent up. The first O.A.O. launchings were highly encouraging.

The full-scale lunar base lies further ahead, but there seems no valid reason why it should not have been set up well before the turn of the century. All branches of science—including medicine—will benefit; and it is to be hoped that the Lunar Base will be established by an international rather than a national programme. A disunited and warring Earth need not necessarily preclude a peaceful Moon.

We cannot be nearly so confident about manned voyages to

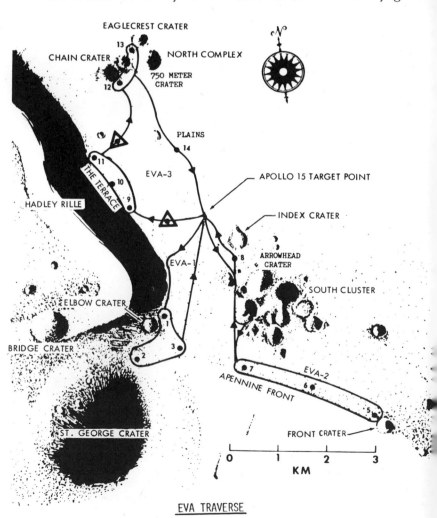

PLANNED EVA FOR APOLLO 15 ASTRONAUTS. *The diagram shows the Extra-Vehicular Activity which was planned for Astronauts Scott and Irwin. In the event, North Complex was not reached, but the astronauts drove to the edge of the great Hadley Rille*

other planets. Chemical-propellent rockets of the Apollo type can hardly be used to take men further than the Moon, and we must await the development of reliable nuclear-powered vehicles. These may well come during the 1980s, and the official American view at the moment is that a Mars expedition will be possible by 1990. Whether or not it will actually be sent must depend partly upon what we learn about Mars in the interim.

Eventually—in two, three, four or more centuries; perhaps much earlier—it will be practicable to send astronauts out to the limits of the Solar System. Whether we shall ever travel further remains to be seen. Material rockets can hardly be used for inter-stellar travel; the time of flight would be much too long, and science-fiction ideas such as space-arks and time-warps are distinctly far-fetched. If we are ever to make direct contact with other beings living on planets of other stars, we must make use of methods which are so completely beyond our understanding as yet that we cannot even speculate about them.

We cannot tell whether an Earthman will ever greet a being from another world, far away from across the Galaxy. And yet whether or not contact is made, we may be sure that such beings exist. It is a sobering thought.

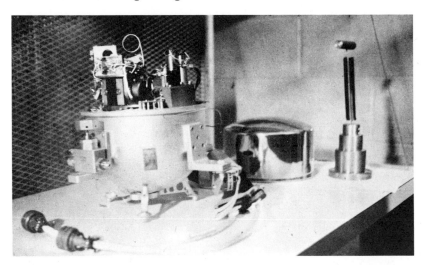

SEISMOMETER *which was taken to the Moon in Apollo 11 in 1969. It was made on the same principle as an Earth seis-mometer, but was more sensitive*

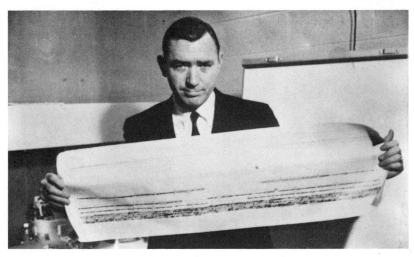

MOONQUAKE TRACE-SIGNALS. *Dr. Gary Lathom, Principal Investigator for the lunar seismic experiment, with the record of the first trace-signals of 'moonquakes' sent back by the Apollo 11 seismometer from the Mare Tranquillitatis. Photograph by Patrick Moore, 1969*

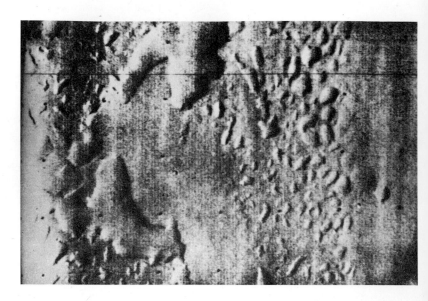

PITS AND HOLLOWS ON MARS, *photographed from Mariner IX in January 1971. The large closed basins to the top are each about 11 miles across. It has been suggested, though not proved, that they are 'polar phenomena', perhaps due to the thawing of ice; the area is about 500 miles from the Martian south pole.*

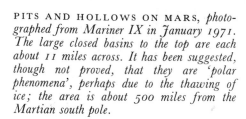

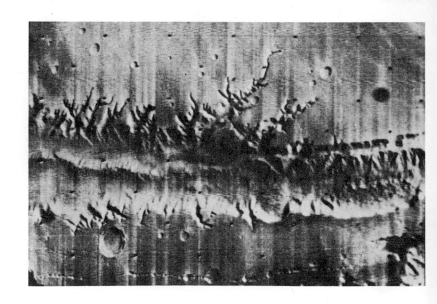

CRATERS ON MARS. *These dark-floored craters in the Phæthontis area of Mars were photographed from Mariner IX. The central crater has a diameter of over 70 miles*

GREAT CHASM ON MARS. *This chasm in Tithonius Lacus, photographed from Mariner IX, is 300 miles long and over 80 miles wide. The 'tributaries' are in fact closed depressions*

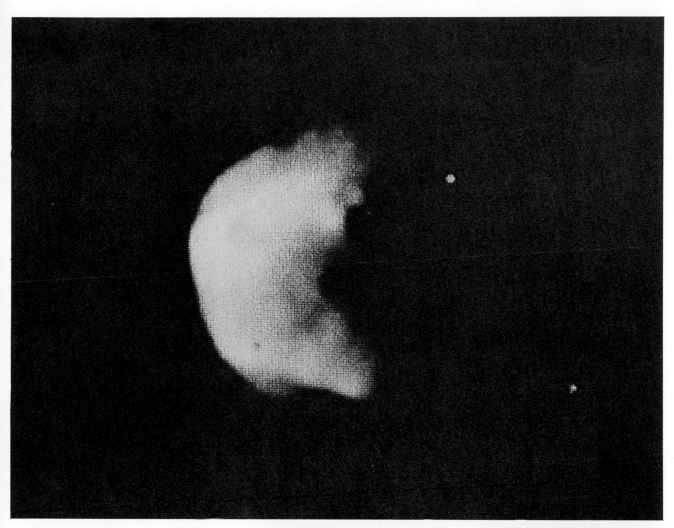

DEIMOS. *The outer satellite of Mars, photographed from Mariner IX. Its greatest diameter is only about 7 miles*

VENERA 7 CAPSULE. *A full-scale model of the capsule which was soft-landed on Venus by Venera 7. The model, on display at the Paris Air Show in 1971, was photographed by Patrick Moore*

'THE CHANDELIER' NEAR NOCTIS LACUS. *Another Mariner IX photograph, showing a great network of canyons on a volcanic landscape*

NOCTIS LACUS. *Here the canyons, photographed from Mariner IX, are over a mile wide. Each is comparable with the Grand Canyon of the Colorado!*

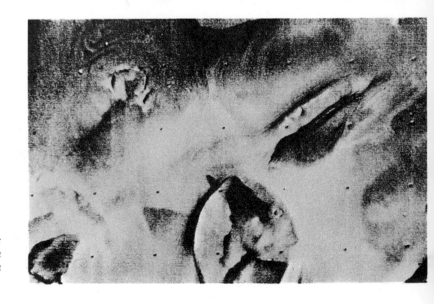

POLAR FEATURES. *Extraordinary pits and grooves not far from the south polar region of Mars (from Mariner IX). The origin of these features is still uncertain*

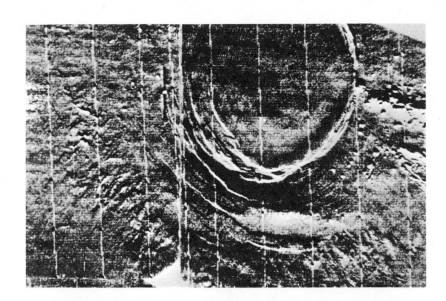

NODUS GORDII. *A shield volcano on Mars, photographed from Mariner IX.*

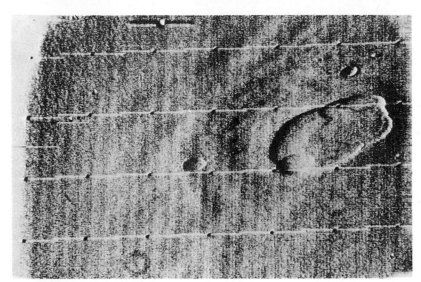

NIX OLYMPICA. *One of the first good photographs of Mars obtained from Mariner IX. The lofty caldera is protruding over the top of the great dust-storm which hid the surface of the planet in the autumn and early winter of 1971*

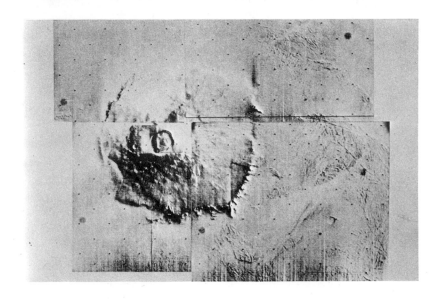

NIX OLYMPICA. *A tremendous volcano of the Hawaiian type—but even larger: the base of the mountain is 300 miles in diameter. Photograph from Mariner IX*

conclusion

WE HAVE SURVEYED the history of astronomy from its dim beginnings up to modern times. Much has been left out, but perhaps enough has been said to show that the tale is indeed a wonderful one.

Two thousand years ago the Earth was believed to be all-important, and to lie in the very centre of the universe. Even 400 years ago, scientists who taught otherwise did so at the peril of their lives. We are wiser now, in this respect at least. We have found that our world is no more important in the cosmos than a single drop of water in the Atlantic; we know that even the Sun is a very junior member of the Galaxy, and that the Galaxy itself is only one among thousands of millions. We have studied star-systems so remote that their light started on its journey towards us before life on Earth began; we have picked up radio waves across vast stretches of space, and we have even sent our pioneers to begin the direct exploration of the Solar System.

If we use our knowledge in the right way, there is no limit to what we can do. Yet we should never forget that we owe everything to those who came before us. Aristarchus . . . Ptolemy . . . Tycho Brahe . . . Kepler . . . Newton . . . Herschel . . . Tsiolkovskii . . . Jansky . . . Gagarin . . . Shepard . . . Armstrong . . . Aldrin . . . What would Tycho, for instance, have thought about the Palomar reflector, the Jodrell Bank 'dish', and men on the Moon? Yet Tycho lived less than four centuries ago. Four centuries hence, men may look back at the pioneers of today, and say the same of them.

The story of astronomy is not over. Indeed, it may only just be starting.

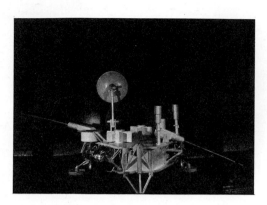

VIKING 'LANDER'. *A full-scale model of the Viking 'lander' which is scheduled to be taken to Mars in 1976. This was on display at the Paris Air Show in 1971, and photographed by Patrick Moore*

RE-USABLE SPACE SHUTTLE. *This is a model of a re-usable space shuttle the United States plans to develop in the seventies. Consisting of a booster and an orbiter stage, the shuttle will take off like a rocket, fly in orbit like a space ship, and land like an airplane. Lines on top of the orbiter show the size of the cargo area that might carry satellites for release in space or transport passenger-scientists to conduct research in space for up to 30 days at a time*

landmarks in the story of astronomy

B.C. (*All dates rather uncertain*)

2000 Constellations first drawn up by the old star-watchers.

580 Pythagoras speculates as to the motion of the Earth and planets.

280 Aristarchus suggests that the Earth moves round the Sun.

270 Eratosthenes measures the size of the Earth.

130 Hipparchus draws up his star catalogue.

A.D.

140 Ptolemy writes his *Almagest*.

813 Al Mamon founds the Baghdad school of astronomy, and has the *Almagest* translated into Arabic.

903 Al-Sûfi draws up his star catalogue.

1054 Supernova in Taurus recorded by Chinese astronomers.

1270 Alphonsine Tables published by order of Alphonso X of Castile.

1433 Ulugh Beigh sets up an observatory at Samarkand.

1440 Nikolaus Krebs (Nicholas of Cusa) speculates about the motion of the Earth.

1474 Regiomontanus suggests the 'lunar distances' method of determining longitude.

1543 Copernicus' *De Revolutionibus Orbium Cœlestium* published.

1572 Tycho Brahe observes a supernova in Cassiopeia.

1576 Tycho founds the observatory at Uraniborg.

1596 Tycho leaves Denmark; Uraniborg abandoned.

1595 Mira Ceti observed by David Fabricius.

1600 Giordano Bruno burned at the stake.

1603 Publication of Johann Bayer's star catalogue, *Uranometria*.

1604 Kepler's supernova in Ophiuchus.

1608 Lippershey develops the telescope.

1609 Galileo first uses the telescope for astronomical purposes.
First two of Kepler's Laws of Planetary Motion published.

1611 Sunspots observed by Galileo, Scheiner and Johann Fabricius.

1612 The Orion Nebula first reported by N. Peiresc.

1618 Kepler's Third Law published.

1627 Kepler publishes the Rudolphine Tables.

1631 Transit of Mercury, predicted by Kepler, observed by Gassendi.

1632 Publication of Galileo's *Dialogue*.

1633 Galileo summoned before the Inquisition at Rome, and forced to recant.

1638 Variability of Mira Ceti discovered by Phocylides Holwarda.

1639 Jeremiah Horrocks observes the transit of Venus.

1647 Publication of Hevelius' map of the Moon.

1651 Publication of Riccioli's map of the Moon.

1655 Huygens discovers Titan, the largest satellite of Saturn, and also the true nature of Saturn's rings.

1656 Founding of the Copenhagen Observatory.

1659 Markings on Mars first seen by Huygens.

1663 Principle of the reflecting telescope put forward by James Gregory.

1665 Newton's experiments on light and gravitation, at Woolsthorpe.

1666 Polar caps on Mars observed by G. D. Cassini.

1668 Newton builds the first reflecting telescope.

1669 Variability of Algol discovered by Montanari.

1671 Paris Observatory founded.
G. D. Cassini discovers Iapetus, the eighth satellite of Saturn.

1675 Greenwich Observatory founded.
G. D. Cassini discovers the main division in Saturn's rings.
O. Rømer measures the velocity of light.

1676 Halley goes to St. Helena to catalogue the southern stars.

1683 Cassini observes the Zodiacal Light.

1687 Publication of Newton's *Principia*.

1704 Publication of Newton's *Opticks*.

1705 Halley predicts the return of Halley's Comet for 1758.

1725 Publication of the final version of Flamsteed's star catalogue.

1728 James Bradley discovers the aberration of light.

1729 Chester More Hall discovers the principle of the achromatic refractor.

1744 Appearance of Chéseaux' six-tailed comet.

1750 Lacaille, at the Cape, catalogues 10,000 stars.

1750 Thomas Wright speculates as to the origin of the Solar System.

1758 Palitzsch discovers Halley's Comet at its predicted return.
Dollond rediscovers the principle of the achromatic refractor.

1761 Transit of Venus observed. Atmosphere of Venus discovered by Lomonosov.

1762 Completion of Bradley's measures of the positions of 60,000 stars.

1767 Nevil Maskelyne founds the *Nautical Almanac*.

1769 Transit of Venus observed. Measures made of the distance between the Earth and the Sun.

1772 Bode's Law publicized by Johann Bode.

1774 First recorded astronomical observation of William Herschel.

1776 Lunar map published by Tobias Mayer.

1779 Johann Schröter founds his observatory at Lilienthal.

1781 Publication of Messier's catalogue of star-clusters and nebulæ.

1781 Herschel discovers Uranus.

1783 Goodricke puts forward his theory of the variability of Algol.

1784 Goodricke discovers the variability of Delta Cephei.

1786 Herschel puts forward his 'disk' theory of the shape of the Galaxy.

1789 Completion of Herschel's 40-foot reflector.

1796 Publication of Laplace's Nebular Hypothesis.

1801 Piazzi discovers Ceres, first of the minor planets.
Publication of Lalande's catalogue of 47,380 stars.

1802 Herschel announces the discovery of binary star systems.
Wollaston observes dark lines in the solar spectrum.

1803 Fall of meteorites at L'Aigle. Nature of meteorites established.

1811 Olbers puts forward his theory of comet-tails.

1813 Destruction of Schröter's observatory at Lilienthal.

1815 Fraunhofer studies the dark lines in the spectrum of the Sun.

1818 Encke's Comet discovered by Pons; return predicted by Encke.

1819 Bessel completes the reduction of Bradley's observations of 5,000 stars.

1824 Erection of Fraunhofer's 'Great Dorpat Refractor'.

1826 Discovery of Biela's Comet, independently by Biela and Gambart.

1832 John Herschel begins his observations of the southern stars, at the Cape.

1833 Great meteor shower (the Leonids).

1834 Bessel discovers the irregularity of the proper motion of Sirius, and attributes it to the presence of a binary companion.

1837 Publication of Beer and Mädler's *Der Mond*, together with the first really accurate map of the Moon.

1838 Bessel measures the distance of a star (61 Cygni).

1840 Draper takes the first photograph of the Moon.

1842 Announcement, by C. Doppler, of Doppler's Principle.

1843 Sunspot cycle discovered by H. Schwabe.

1845 Completion of Lord Rosse's 72-inch reflector at Birr Castle.
Hencke discovers the fifth minor planet, Astræa.
Sun photographs taken by Fizeau and Foucault.

1845 Breaking-up of Biela's Comet observed.

1846 Discovery of Neptune by Adams and Le Verrier.

1847 15-inch refractor set up at Cambridge, Massachusetts, U.S.A.

1848 Roche proves that Saturn's rings cannot be solid (Roche's Limit).

1850 Crêpe Ring of Saturn discovered by Bond.

1854 The Gegenschein discovered by the Danish astronomer T. Brorsen.

1858 Appearance of Donati's Comet.

1859 Kirchhoff interprets the dark lines in the spectra of the Sun and stars.

1861 Announcement of Spörer's law of sunspots.

1862 Sirius B discovered by Clark, in the position forecast by Bessel.

1863 Secchi first classifies the stars into special types.
Huggins identifies elements in the spectra of Betelgeux and Aldebaran.

1864 Huggins proves the gaseous nature of the irresolvable nebulæ.

1865 Publication of Jules Verne's story *From the Earth to the Moon*.

1867 Wolf-Rayet stars first described by G. Wolf and G. A. Rayet.

1868 Solar prominences observed by Jansen and Lockyer, without waiting for an eclipse.
Ångström publishes an accurate map of the solar spectrum.

1872 The Bieliid meteor shower.

1874 Founding of the Observatory at Meudon (France).
Transit of Venus observed. Astronomical unit re-measured.

1877 Schiaparelli describes the Martian canals, and A. Hall discovers Phobos and Deimos.

1878 Publication of Julius Schmidt's map of the Moon.
The Great Red Spot on Jupiter becomes very conspicuous.

1882 Transit of Venus. New measures of the astronomical unit.

1890 Announcement of Lockyer's theory of stellar evolution.
E. C. Pickering gives a more detailed classification of stellar spectra.
H. Vogel establishes the existence of spectroscopic binaries.

1891 The spectroheliograph invented by Hale, and independently by Deslandres.

1894 Flagstaff Observatory, in Arizona, founded by Percival Lowell.

1896 33-inch refractor erected at Meudon.

1897 Yerkes Observatory opened.

1900 Lebedev establishes the reality of light-pressure.
New theory of the origin of the Solar System put forward by Chamberlin and Moulton.

1901 Appearance of Nova Persei.

1903 Publication of Tsiolkovskii's first paper on astronautics.

1905 Mount Wilson Observatory founded.

1908 E. Hertzsprung describes giant and dwarf stars.

1908 The 60-inch reflector erected at Mount Wilson.
Fall of the Siberian Meteorite.

1912 Miss Leavitt's studies of Cepheids lead to the discovery of the Period-Luminosity Law.

1913 H. N. Russell puts forward his theory of stellar evolution.

1914 Cepheid pulsation studied by H. Shapley.
Goddard begins practical experiments with rockets.

1915 W. S. Adams studies the spectrum of Sirius B, which leads to the discovery of White Dwarf stars.

1917 Completion of the 100-inch Hooker reflector at Mount Wilson.

1918 Shapley's studies leading to the first accurate idea of the shape of the Galaxy.

1919 Important catalogue of dark nebulæ published by E. E. Barnard.

1920 The Red Shifts in the spectra of galaxies announced by V. M. Slipher.

1923 Hubble proves that the galaxies lie beyond the Milky Way.
Hale invents the spectrohelioscope.
Publication of H. Oberth's book about astronautics.

1926 Goddard fires the first liquid-fuel rocket.

1927 Studies by J. H. Oort show that the centre of the Galaxy lies in the direction of the Sagittarius star-clouds.

1930 Pluto discovered by C. Tombaugh, from Lowell's calculations.
B. Schmidt invents the Schmidt Camera.

1931 Close approach of Eros to the Earth. Astronomical unit re-measured.
K. Jansky discovers radio waves coming from the Milky Way.

1932 Dunham discovers carbon dioxide in the atmosphere of Venus.

1934 White spot discovered on Saturn.
Appearance of a nova in Hercules, discovered by J. Prentice.

1937 Grote Reber builds his 'dish' radio telescope.

1938 New theory of stellar energy proposed by H. Bethe, and independently by C. Von Weizsäcker.

1942 K. A. Strand announces that 61 Cygni B is attended by a planet.

1944 H. van de Hulst suggests that interstellar hydrogen must emit radio waves at 21·1 cm.

1946 Radar echoes from the Moon recorded by Z. Bay.

1948 Completion of the Hale 200-inch reflector at Palomar.

1949 The first step-rocket fired at White Sands, New Mexico.

1951 H. Ewen and E. Purcell discover the 21·1-cm. emission predicted by van de Hulst.

1952 Baade announces the revision of the distance-scale of the galaxies.

1955 Completion of the 250-foot radio telescope at Jodrell Bank.

1957 The first artificial satellites are launched by the Russians.

1958 The first American artificial satellites launched.
N. Kozirev observes an outbreak inside the lunar crater Alphonsus.

1959 The Russian Luniks: Lunik I (past the Moon), Lunik II (landing on the Moon), Lunik III (sending back photographs of the far side of the Moon).

1960 Further launchings of artificial satellites.

1961 Soviet Venus probe launched.
First orbital flight, by Y. Gagarin.
Sub-orbital flight, by A. Shepard.

1962 First American orbital flight, by J. Glenn.
Further orbital flights by Russians and Americans.
Further planetary probes: Mars I (Russia) and Mariner II (U.S.A.)

1963 New measures of the astronomical unit.
Van de Kamp reports a planet associated with Barnard's Star.
Quasars found to be outside our Galaxy.

1964 Close-range pictures of the Moon from Ranger VII (U.S.A.).

1965 Improved close-range lunar photographs from Rangers VIII and IX (U.S.A.).
Improved photographs of the Moon's far side from Zond III (Russia).
Close-range photographs of Mars from Mariner IV (U.S.A.).

1966 First soft landing on the Moon: Luna 9 (Russia).
Russian probe lands on Venus.
Luna 10 (Russia) put into an orbit round the Moon.
First soft landing of an American probe on the Moon (Surveyor I).
First photographs from the American circum-lunar probe Orbiter.

1967 Measures of the Moon's surface density made from Luna 13.
First soft landing by an automatic probe on Venus (Venera 4, Russia).
Discovery of pulsars, by Jocelyn Bell at Cambridge University.

1968 First manned flight round the Moon: Borman, Lovell and Anders in Apollo 8 (U.S.A.).

1969 First men on the Moon: Armstrong and Aldrin, from Apollo 11 (U.S.A.).
Improved photographs and data from Mars obtained by Mariners VI and VII (U.S.A.).

1970 First successful soft landing on Venus: Venera 7 (Russia).

1971 Final testing of the mounting for the new 236-inch Soviet reflector.
First manned mechanical vehicle on the Moon: Apollo 15.
First probes in orbit round Mars.
First soft-landing on Mars (Mars 3, Russia).

1972 Launch of Pioneer 10, first probe to Jupiter (U.S.A.).
Apollo 16 (U.S.A.).

list of illustrations

● Indicates four-color illustrations

acknowledgments

Illustrations are keyed by the appropriate page number followed by positioning symbol U for Upper, C for Centre, L for Lower, l for Left, r for Right

Macdonald & Co. (Publishers) Ltd; 13L, 14, 15L, 16, 18, 19, 20L, 22U, 22L, 23, 24, 25U, 28U, 28L, 29U, 29L, 32, 33U, 25U, 28U, 38L, 29U, 40U, 41, 42U, 42C, 42L, 43U, 43L, 44U, 45U, 45L, 46U, 46L, 49U, 50U, 50L, 51L, 55U, 56, 60U, 60L, 61, 62L, 65, 66U, 67U, 67L, 69L, 72L, 73, 74U, 74L, 75, 81L, 82L, 83L, 84U, 85U, 85L, 86L, 88U, 89, 91L, 92L, 95U, 97U, 98, 99U, 100U, 100C, 100L, 101L, 102U, 102L, 103, 104L, 105, 122U, 126, 127, 134L, 136U, 137U, 137L, 138U, 138C, 140, 141, 142L, 144, 145, 146U, 147L, 148U, 149, 150U, 150C, 153L, 154U, 154L, 155U, 155L, 157L, 159, 160L, 161U, 161L, 162U, 162L, 169U, 171, 172L, 173U, 174, 175, 177L, 179, 180, 183U, 184U, 184L, 185U, 186C, 188, 189, 190U, 190L, 191U, 191C, 196U, 206U, 207U, 207L, 214C, 215C, 224, 228U, 228L.

Mount Wilson and Palomar Observatories; 12, 13U, 17, 48C, 57, 58, 59Ur, 66L, 71L, 82U, 94L, 97L, 101U, 108, 111C, 114U, 114C, 114L, 115, 118U, 135L, 146L, 150L, 151U, 160U, 164U, 164L, 167U, 168L, 169L 170, 176, 178, 183C, 183L, 185L, 186L, 187U, 187L, 191L, 192, 193U, 193L, 194U, 194L, 195U, 195L, 196L, 197U, 197L, 198U, 198L, 199L, 200U, 200L, 203U, 203L, 206L, 210L, 217C, 218U, 218L, 219L, 223U.

Allan Lanham; 15U.

Daily Express; 20U.

F. J. Acfield, Forest Hall Observatory, Northumberland; 21U.

H. E. Wood, Union Observatory, Johannesburg; 21L.

Radio Times and Hulton Picture Library; 26U, 31L, 33L, 47U, 48U.

Deutsches Museum, Munich; 26L, 104U.

Crown Copyright, Science Museum, London; 27U, 34L, 71U, 88U, 90L, 107U, 111U.

Science Museum, London; 27L, 34U, 35L, 36U, 36L, 37U, 54U, 54L, 62U, 64U, 77L, 79U, 95L, 157C.

W. T. O'Dea; 30U, 30L, 31U.

Gösta Persson; 37L.

Copyright by the California Institute of Technology and Carnegie Institute of Washington; 44L, 93, 146C, 146L, 201, 202U, 202L, 204U, 204L, 205U, 205L, 216.

Patrick Moore; 49L, 53U, 69U, 70U, 72U, 77U, 78L, 79L, 84L, 86U, 87L, 90U, 94U, 96U, 96L, 108U, 109, 110L, 112U, 112L, 113U, 115L, 117L, 118L, 119U, 119C, 120L, 130L, 134U, 139L, 156C, 158U, 209L, 211L, 213U, 213C, 213L, 237L, 239L, 242U.

Dominic Fidler; 47L.

Lowell Observatory; 51U, 64L, 80, 106U, 106L, 147C, 220C.

Lick Observatory; 51C, 63U, 63L, 70L, 112C, 121U, 132U.

Father D. J. K. O'Connell, S.J., Vatican Observatory; 52U, 52L.

W. M. Baxter; 53L, 55L, 59Ul, 135U, 165U, 165L.

Eastern Daily Press; 59L.

Royal Astronomical Society; 81U, 133C, 199U, 199C, 199Lr.

K. S. G. Stocker; 68Ul, 68Ur, 68L.

British Museum; 76.

National Maritime Museum, Greenwich; 78U.

B. Warner and **T. Salmundsson,** University of London Observatory; 91U.

H. R. Hatfield; 92U, 186U.

H. P. Wilkins; 99L.

Science Museum, London; Reproduced by courtesy of the Royal Society, 107L.

Port Elizabeth Observatory, South Africa; 110U.

Colin Ronan; 113, 209Ul, 230U.

U.S. Naval Observatory, Washington; 116.

Australian News and Information Bureau; 117, 119L, 211U, 214U, 215L, 219U, 232L.

Brian Gulley; 120U.

Pic du Midi Observatory; 121L, 122L, 124Cr, 125L.

National Aeronautics and Space Administration, Washington; 123U, 123L, 124C, 124L, 125U, 125L, 128U, 128C, 128L, 129U, 129C, 130U, 130C, 131U, 131C, 131L, 132U, 132Lr, 143U, 221U, 221L, 226, 227, 229, 231, 235U, 236, 237U, 239U, 240U, 240C, 240L, 241U, 241C, 241L. Endpapers.

Novosti Press Agency; 129L.

U.S. Embassy, London; 132C, 132L, 143C, 143L, 225l, 225r, 233L, 238U, 238C, 238L, 242L.

G. V. Schiaparelli; 133U, 220U.

H. E. Dall; 136L.

Ramon Lane; 139U.

W. T. Hay; 147U.

E. A. Whitaker; 151L.

J. Bennett; 152.

G. E. D. Alcock; 153U, 153C.

H. B. Ridley; 156U, 156L, 157U, 173L.

D. E. Blackwell and **M. F. Ingham;** 158C.

G. J. H. McCall; 158L.

H. Brinton; 163.

P-A Reuter; 166.

Professor Ake Wallenquist; 167L.

Royal Greenwich Observatory; 168U.

Elliot and Fry Ltd; 177U.

W. S. Franks; 182.

Royal Observatory, Cape; 196C.

Central Office of Information; 208, 212, 233U.

McDonnell Aircraft Corporation; 209Ur.

Karl Jansky; 210U.

Paris Observatory; 220L.

National Radio Astronomy Observatory, Green Bank; 223L.

National Film Board of Canada; 230Ll, 230Lr.

U.S. Army Air Force; 232U.

K. Gottlieb, Mount Stromlo Observatory, Canberra; 233C.

Soviet Weekly; 234U, 234L, 235U, 238L, 239U.

Photocraft Picture by Morris Allan, Dunfermline; 234L.

index